Springer Series in
OPTICAL SCIENCES 155

founded by H.K.V. Lotsch

Springer Series in
OPTICAL SCIENCES

The Springer Series in Optical Sciences, under the leadership of Editor-in-Chief *William T. Rhodes*, Georgia Institute of Technology, USA, provides an expanding selection of research monographs in all major areas of optics: lasers and quantum optics, ultrafast phenomena, optical spectroscopy techniques, optoelectronics, quantum information, information optics, applied laser technology, industrial applications, and other topics of contemporary interest.

With this broad coverage of topics, the series is of use to all research scientists and engineers who need up-to-date reference books.

The editors encourage prospective authors to correspond with them in advance of submitting a manuscript. Submission of manuscripts should be made to the Editor-in-Chief or one of the Editors. See also www.springer.com/series/624

Editor-in-Chief
William T. Rhodes
Georgia Institute of Technology
School of Electrical and Computer Engineering
Atlanta, GA 30332-0250, USA
E-mail: bill.rhodes@ece.gatech.edu

Editorial Board
Ali Adibi
Georgia Institute of Technology
School of Electrical and Computer Engineering
Atlanta, GA 30332-0250, USA
E-mail: adibi@ee.gatech.edu

Toshimitsu Asakura
Hokkai-Gakuen University
Faculty of Engineering
1-1, Minami-26, Nishi 11, Chuo-ku
Sapporo, Hokkaido 064-0926, Japan
E-mail: asakura@eli.hokkai-s-u.ac.jp

Theodor W. Hänsch
Max-Planck-Institut für Quantenoptik
Hans-Kopfermann-Straße 1
85748 Garching, Germany
E-mail: t.w.haensch@physik.uni-muenchen.de

Takeshi Kamiya
Ministry of Education, Culture, Sports
Science and Technology
National Institution for Academic Degrees
3-29-1 Otsuka, Bunkyo-ku
Tokyo 112-0012, Japan
E-mail: kamiyatk@niad.ac.jp

Ferenc Krausz
Ludwig-Maximilians-Universität München
Lehrstuhl für Experimentelle Physik
Am Coulombwall 1
85748 Garching, Germany   *and*
Max-Planck-Institut für Quantenoptik
Hans-Kopfermann-Straße 1
85748 Garching, Germany
E-mail: ferenc.krausz@mpq.mpg.de

Bo Monemar
Department of Physics
and Measurement Technology
Materials Science Division
Linköping University
58183 Linköping, Sweden
E-mail: bom@ifm.liu.se

Herbert Venghaus
Fraunhofer Institut für Nachrichtentechnik
Heinrich-Hertz-Institut
Einsteinufer 37
10587 Berlin, Germany
E-mail: venghaus@hhi.de

Horst Weber
Technische Universität Berlin
Optisches Institut
Straße des 17. Juni 135
10623 Berlin, Germany
E-mail: weber@physik.tu-berlin.de

Harald Weinfurter
Ludwig-Maximilians-Universität München
Sektion Physik
Schellingstraße 4/III
80799 München, Germany
E-mail: harald.weinfurter@physik.uni-muenchen.de

Please view available titles in *Springer Series in Optical Sciences*
on series homepage http://www.springer.com/series/624

Motoichi Ohtsu

Editor

Progress
in Nano-Electro-Optics VII

Chemical, Biological,
and Nanophotonic Technologies
for Nano-Optical Devices and Systems

With 79 Figures

 Springer

Editor
Prof. Dr. Motoichi Ohtsu
Department of Electronics Engineering
School of Engineering
The Universit of Tokyo
7-3-1 Hongo, Bunkyo-ku, Tokyo 113-8656, Japan
E-mail: ohtsu@ee.t.u-tokyo.ac.jp

ISSN 0342-4111   e-ISSN 1556-1534
ISBN 978-3-642-03950-8   e-ISBN 978-3-642-03951-5
DOI 10.1007/978-3-642-03951-5
Springer Heidelberg Dordrecht London New York

Library of Congress Cataloging-in-Publication Data

Progress in nano-electro-optics VII : Chemical, Biological, and nanophotonic technologies for nano-optical devices and systems / Motoichi Ohtsu (ed.). p.cm. – (Springer series in optical sciences ; v. 155)
Includes bibliographical references and index.
ISBN 978-3-642-03950-8 (alk. paper)
1. Electrooptics. 2. Nanotechnology. 3. Near-field microscopy. I. Ohtsu, Motoichi. II. Series.
TA1750 .P75 2002   621.381'045–dc21   200203032

Typesetting by the authors and Integra, using a Springer LaTeX macro
Cover design: eStudio Calamar Steinen

Printed on acid-free paper

Springer is part of Springer Science+Business Media (www.springer.com)

Preface to *Progress in Nano-Electro-Optics*

Recent advances in electro-optical systems demand drastic increases in the degree of integration of photonic and electronic devices for large-capacity and ultrahigh-speed signal transmission and information processing. Device size has to be scaled down to nanometric dimensions to meet this requirement, which will become even more strict in the future. In the case of photonic devices, this requirement cannot be met only by decreasing the sizes of materials. It is indispensable to decrease the size of the electromagnetic field used as a carrier for signal transmission. Such a decrease in the size of the electromagnetic field beyond the diffraction limit of the propagating field can be realized in optical near fields.

Near-field optics has progressed rapidly in elucidating the science and technology of such fields. Exploiting an essential feature of optical near fields, i.e., the resonant interaction between electromagnetic fields and matter in nanometric regions, important applications and new directions such as studies in spatially resolved spectroscopy, nano-fabrication, nano-photonic devices, ultrahigh-density optical memory, and atom manipulation have been realized and significant progress has been reported. Since nano-technology for fabricating nanometric materials has progressed simultaneously, combining the products of these studies can open new fields to meet the above-described requirements of future technologies.

This unique monograph series entitled "Progress in Nano-Electro-Optics" is being introduced to review the results of advanced studies in the field of electro-optics at nanometric scales and covers the most recent topics of theoretical and experimental interest on relevant fields of study (e.g., classical and quantum optics, organic and inorganic material science and technology, surface science, spectroscopy, atom manipulation, photonics, and electronics). Each chapter is written by leading scientists in the relevant field. Thus, high-quality scientific and technical information is provided to scientists, engineers, and students who are and will be engaged in nano-electro-optics and nano-photonics research.

I gratefully thank the members of the editorial advisory board for valuable suggestions and comments on organizing this monograph series. I wish to express my special thanks to Dr. T. Asakura, Editor of the Springer Series in Optical Sciences, Professor Emeritus, Hokkaido University for recommending me to publish this monograph series. Finally, I extend an acknowledgement to Dr. Claus Ascheron of Springer-Verlag, for his guidance and suggestions, and to Dr. H. Ito, an associate editor, for his assistance throughout the preparation of this monograph series.

Yokohama, October 2002   *Motoichi Ohtsu*

Preface to Volume VII

This volume contains five review articles focusing various, but mutually related topics in nano electro-optics. The first article describes recent developments in the study of the temperature-induced phase transition and photo-induced phase transition of ferromagnetic RbMnFe complex. In addition, with non phase transition material of RbMnFe, the light-induced phase collapse is demonstrated, which may provide a good strategy for the next generation high density optical recording. As photo-induced phase transition at room temperature, large yield and fast response of the photo-conversion from the low- to high-temperature phase will allow us to consider a new type of optical switching device.

The second article is devoted to describing recent achievements relating to photo-induced energy transfer in artificial photosynthesis. In particular, the emphasis lies on self-assembled multi-porphyrin array that are highly promising materials for photo-catalysts, organic solar cells, and molecular optoelectronic devices. Well-defined molecule-based nanoarchitectures exhibiting energy transfer will open the door to nanoscience and nanotechnology, which stimulates a variety of fields including chemistry, biology, physics, and electronics to develop new scientific and technological principles and concepts.

The third article concerns the homoepitaxial growth and multiple-quantum wells (MQW) in ZnO. Fabrications on MQWs and their low-dimensional optical properties are discussed. Self-organized surface nanowires on M-nonpolar ZnO layers are also described, wherein discussions concentrate on a growth mechanism and developments concerning low-dimensional structures of quantum wires. Further, discussions of various properties of ZnCoO diluted magnetic semiconductors and fabrications of the quantum wells geometries are also given. Demonstrated homoepitaxial technique can be effective for electro-, magneto-, and optical applications based on ZnO.

The fourth article deals with two topics. The first topics is a novel polishing technique that uses near-field photochemical etching based on a nonadiabatic process, with which the roughness of an ultra-flat silica surface can be reduced to an Angstrom- level. Since this technique is a noncontact method without a polishing pad, it can be applied not only to flat surfaces but also to three-dimensional surfaces. Furthermore, this method is also compatible with mass production. The second topics is the recent achievements with nanophotonics devices based on spherical quantum dots. Optical near-field energy transfer is described.

The last article describes polarization control in the optical near-field and far-field by designing the shape of metal nanostructure. Its application to multi-layer structures and optical security are also discussed. In particular the problem on rotating the plane of polarization is discussed, which should be solved for various applications; devices including nanostructures have already been employed, for instance, in so-called wire-grid polarizers.

As was the case of Vols. I–VI, this volume is published by the supports of an associate editor and members of editorial advisory board. They are:

Associate editor: Yatsui, T. (Univ. of Tokyo, Japan)
Editorial advisory board: Barbara, P.F. (Univ. of Texas, USA)
                          Bernt, R. (Univ. of Kiel, Germany)
                          Courjon, D. (Univ. de Franche-Comté, France)
                          Hori, H. (Univ. of Yamanashi, Japan)
                          Kawata, S. (Osaka Univ., Japan)
                          Pohl, D. (Univ. of Basel, Switzerland)
                          Tsukada, M. (Tohoku Univ., Japan)
                          Zhu, X. (Peking Univ., China)

I hope that this volume will be a valuable resource for the readers and future specialists.

Tokyo, June 2009   *Motoichi Ohtsu*

Contents

List of Contributors

Hirokazu Hori
Interdisciplinary Graduate School of
Medicine and Engineering
University of Yamanashi
4-3-11 Takeda, Kofu
Yamanashi 400-8511, Japan
hirohori@yamanashi.ac.jp

Hiroshi Imahori
Institute for Integrated Cell-Material
Sciences (iCeMS)
Kyoto University
Nishikyo-ku, Kyoto 615-8510, Japan
and
Department of Molecular En-
gineering, Graduate School of
Engineering
Kyoto University
Nishikyo-ku, Kyoto 615-8510, Japan
and
Fukui Institute for Fundamental
Chemistry
Kyoto University
34-4, Takano-Nishihiraki-cho,
Sakyo-ku, Kyoto 606-8103, Japan
imahori@scl.kyoto-u.ac.jp

Tadashi Kawazoe
School of Engineering
The University of Tokyo
2-11-16 Yayoi, Bunkyo-ku
Tokyo 113-8656, Japan
kawazoe@ee.t.u-tokyo.ac.jp

Hiroaki Matsui
Graduate School of Engineering, Department of Electronic
Engineering
The University of Tokyo
7-3-1 Hongo, Bunkyo-ku, Tokyo
113-8656, Japan
hiroaki@ee.t.u-tokyo.ac.jp

Makoto Naruse
New Generation Network Research
Center
National Institute of Information
and Communications Technology
4-2-1 Nukui-kita, Koganei
Tokyo 184-8795, Japan
and
School of Engineering
The University of Tokyo
2-11-16 Yayoi, Bunkyo-ku
Tokyo 113-8656, Japan
naruse@nict.go.jp

Wataru Nomura
School of Engineering
The University of Tokyo
2-11-16 Yayoi, Bunkyo-ku
Tokyo 113-8656, Japan
nomura@nanophotonics.t.u-tokyo.ac.jp

Shin-ichi Ohkoshi
Department of Chemistry, School of Science
The University of Tokyo,
7-3-1 Hongo, Bunkyo-ku, Tokyo,
113-0033 Japan
ohkoshi@chem.s.u-tokyo.ac.jp

Motoichi Ohtsu
School of Engineering
The University of Tokyo
2-11-16 Yayoi, Bunkyo-ku
Tokyo 113-8656, Japan
ohtsu@ee.t.u-tokyo.ac.jp

Hitoshi Tabata
Graduate School of Engineering, Department of Electronic Engineering
The University of Tokyo
7-3-1 Hongo, Bunkyo-ku, Tokyo
113-8656, Japan
tabata@bioeng.t.u-tokyo.ac.jp

Naoya Tate
School of Engineering
The University of Tokyo
2-11-16 Yayoi, Bunkyo-ku
Tokyo 113-8656, Japan
tate@nanophotonics.t.u-tokyo.ac.jp

Hiroko Tokoro
Department of Chemistry, School of Science
The University of Tokyo
7-3-1 Hongo, Bunkyo-ku, Tokyo,
113-0033 Japan
and
PRESTO, JST, 4-1-8 Honcho
Kawaguchi, Saitama, 332-0012 Japan
tokoro@light.t.u-tokyo.ac.jp

Tomokazu Umeyama
Department of Molecular Engineering, Graduate School of Engineering
Kyoto University
Nishikyo-ku, Kyoto 615-8510, Japan
umeyama@scl.kyoto-u.ac.jp

Takashi Yatsui
School of Engineering
The University of Tokyo
2-11-16 Yayoi, Bunkyo-ku
Tokyo 113-8656, Japan
yatsui@ee.t.u-tokyo.ac.jp

Photo-Induced Phase Transition in RbMnFe Prussian Blue Analog-Based Magnet

H. Tokoro and S. Ohkoshi

1.1 Introduction

Studies that are related to temperature-induced phase transitions and photo-induced phase transitions are extensively investigated in solid-state chemistry [1–4]. Temperature-induced phase transition phenomena are observed in spin crossover or intramolecular electron transfer. In a spin crossover complex, a transition metal ion can be in either the low-spin or the high-spin state depending on the strength of the ligand field. When the thermal energy is close to the exchange energy that corresponds to the crossover, a spin transition occurs between the two spin states. This phenomenon is observed in octahedral coordinate iron transition metal complexes [5–7]. Charge-transfer phase transitions have been observed in mixed-valence complexes [7–13], e.g., $[M^{III}_2 M^{II} O(O_2 C_2 H_3)_6 L_3]$ (M= Fe, Mn; L=H_2O, pyridine) [12] and $M(dta)_4 I$ (M= Ni, Pt; dta= dithioacetato) [13]. Charge-transfer phase transitions that accompany spin crossovers have also been reported, e.g., $Co(py_2 X)(3,6-DBQ)_2$ (X= O, S, Se) [14] and $Na_{0.4} Co_{1.3}[Fe(CN)_6] \cdot 4.9 H_2 O$ [15]. A temperature-induced phase transition often accompanies a thermal hysteresis loop, which is related to the cooperativity of the corresponding system. The cooperativity in a metal complex assembly is due to the interaction between a metal ion and lattice strain, e.g., an electron-phonon coupling [16], a Jahn–Teller distortion [17], and an elastic interaction [18]. Cyano-bridged metal assemblies such as hexacyanometalate- [4, 19–38] and octacyanometalate-based magnets [39–44] are suitable for observing a thermal phase transition since they are mixed-valence compounds that have a strong cooperativity due to the CN ligand bridges.

To date, several types of photo-induced phase-transition phenomena have been reported, for example, a light-induced crystalline-amorphous transformation in chalcogenide material [45–47], a light-induced spin-state change on the transition metal ion of a metal complex [3, 48, 49], a light-induced charge transfer in donor-accepter stacked molecules [2, 50, 51], ferromagnetic bimetallic assemblies [35–44], or perovskite manganite [52, 53]. Until now, we have demonstrated photomagnetic effects such as photo-induced magnetization and the photo-induced magnetic pole inversion with cyano-briged bimetallic assemblies [4, 35, 38–44]. One possible method for achieving optical control of magnetization is to change the electron spin state of a magnetic material. For example, if photo-irradiation varies the oxidation numbers of transition metal ions within a magnetic material, its magnetization will be controlled. The bistability of the electronic states is also indispensable for observing photo-induced persistent magnetization since the energy barrier between these bistable states can maintain the photo-produced state even after photo-irradiation is ceased.

From this viewpoint, Prussian blue analogs are an attractive system due to their high T_c values [22]. In particular, Verdaguer et al. reported that $V^{II}[Cr^{III}(CN)]_{0.86} \cdot 2.8H_2O$ exhibits a T_c value of 315 K [24]. Successively, Girolami et al. and Miller et al. reported crystalline $K^I V^{II}[Cr^{III}(CN)_6]$ with $T_c = 103°C$ and amorphous $K^I_{0.058} V^{II/III}[Cr^{III}(CN)_6]_{0.79}(SO_4)_{0.058} \cdot 0.93H_2O$ with $T_c = 99°C$ powder, respectively [27, 28]. In multi-metal Prussian blue analogs, the rational design of magnets based on the molecular field theory is possible for the following reasons: (1) metal substitutions induce only small changes in the lattice constant and (2) superexchange interactions are only essentially effective between the nearest neighbor metal ions [25]. For example, we have designed a novel type of magnet that exhibits two compensation temperatures with the system of $(Ni^{II}_{0.22}Mn^{II}_{0.60}Fe^{II}_{0.18})_{1.5}[Cr^{III}(CN)_6] \cdot 7.5H_2O$ i.e., the spontaneous magnetization changes sign twice as the temperature is varied [29]. In this study, we show the temperature-induced phase transition and photo-induced phase transition of ferromagnetic $Rb_x Mn[Fe(CN)_6]_{(x+2)/3} \cdot zH_2O$ complex.

1.2 Synthesis of Rubidium Manganese Hexacyanoferrate

Preparing method of rubidium manganese hexacyanoferrate, $Rb_x Mn[Fe(CN)_6]_{(x+2)/3} \cdot zH_2O$, is as follows: an aqueous solution (0.1 mol dm^{-3}) of $Mn^{II}Cl_2$ with a mixed aqueous solution of Rb^ICl (1 mol dm^{-3}) and $K_3[Fe^{III}(CN)_6]$ (0.1 mol dm^{-3}) was reacted to yield a precipitate. The precipitate was filtered, dried, and yielded a powdered sample. The prepared compound was a light brown and elemental analyses for Rb, Mn, and Fe indicated that the obtained precipitate had a formula of $RbMn[Fe(CN)_6]$ ($x = 1$, $z = 0$). The 1: 1: 1 ratio of Rb: Mn: Fe allowed the Mn i ons to coordinate six cyanonitrogens. Consequently, the network does not contain water molecules. Scanning electron microscope (SEM) images showed that the obtained powdered sample was composed of cubic microcrystals that were 2.1 ± 1.1 m. For the sample of different x, the sample was prepared by reacting an aqueous solution (0.1 mol dm^{-3}) of $Mn^{II}Cl_2$ with a mixed aqueous solution of Rb^IC (0.05–1 mod dm^{-3}) and $K_3[Fe^{III}(CN)_6]$ (0.1 mol dm^{-3}). The schematic structure of $Rb_x Mn[Fe(CN)_6]_{(x+2)/3} \cdot zH_2O$ is shown in Fig. 1.1.

(a)   (b)

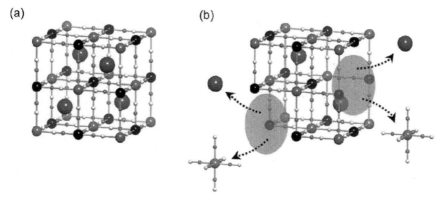

Fig. 1.1. Schematic structures of **(a)** $Rb^I Mn^{II} [Fe^{III}](CN)_6$ and **(b)** $Rb_x^I Mn^{II}$ $[Fe^{III}(CN)_6]_{(x+2)/3} \cdot zH_2O$. *Large gray circle* is Rb^I ion, *middle black circle* is Mn^{II} ion, *middle gray circle* is Fe^{III} ion, *small gray circle* is C atom, and *small white circle* is N atom, respectively. *Shadows* indicate defects at the $Fe^{III}(CN)_6$ sites. Water molecules are omitted for clarity

1.3 Crystal Structure of Rubidium Manganese Hexacyanoferrate

To study the crystal structure of rubidium manganese hexacyanoferrate, X-ray single crystal analysis was performed for $Rb_{0.61}Mn[Fe(CN)_6]_{0.87} \cdot 1.7H_2O$ [54]. Crystal was obtained by the slow diffusion of $MnCl_2$ (7×10^{-3} mol dm^{-3}) dissolved in ethanol into $K_3[Fe(CN)_6]$ (3×10^{-3}mol dm^{-3}) and RbCl (1.4×10^{-2} mol dm^{-3}) dissolved in water for three month. The obtained single crystals measured approximately $0.1 \times 0.1 \times 0.05$ mm^3. Elemental analysis of Rb, Mn, and Fe of the single crystal was performed by microscopic fluorescent X-ray analysis (micro-FXA) with an X-ray spot size of φ 10 μm. The observed ratio of metal ions was Rb:Mn:Fe = 0.58($\pm$ 0.04) : 1.00($\pm$ 0.03) : 0.86($\pm$ 0.03). The density (d) measured by the flotation method (tetrabromoethane and toluene) showed $d = 1.84(3)$ g cm^3. These results of micro-FXA and density measurements showed that the formula of the crystal was $Rb^I_{0.61}Mn^{II}[Fe^{III}(CN)_6]_{0.87} \cdot 1.7H_2O$. Crystal data was collected on a Rigaku RAXIS RAPID imaging plate area detector with graphite monochromated Mo-Ka radiation.

The present single crystal , $Rb_{0.61}Mn[Fe(CN)_6]_{0.87} \cdot 1.7H_2O$, contains an intermediate composition value of 0.61 for Rb^+. This compound has vacancies of $0.13 \times [Fe(CN)_6]$ in the cubic lattice to maintain charge neutrality . It is expected that the Mn ion around the vacancy is coordinated to a water molecule (so-called ligand water) and the interstitial sites are occupied by Rb ions or non-coordinated waters (so-called zeolitic water molecules).

X-ray crystallography shows that $Rb^I_{0.61}Mn^{II}[Fe^{III}(CN)_6]_{0.87} \cdot 1.7H_2O$ belongs to the face-centered cubic lattice Fm$\bar{3}$m with lattice constants of

Table 1.1. Crystallographic and refinement data

Formula	$Rb_{0.61}Mn[Fe(CN)_6]_{0.87}\cdot1.7H_2O$
Fw	322.1
Calculated density / g cm^{-3}	1.829
Temperature / K	93.1
Crystal system	Cubic
Space group	$Fm\,3m$
Lattice constants / Å	10.5354(4)
Unit cell volume / Å^3	1169.37(8)
Number of formula units Z	4
Absorption coefficient μ / cm^{-1}	46.66
Number of measured reflections	20363
Number of independent reflections	292
Number of refined parameters	15
GOF on $|F|^2$	1.270
$R1\,[I > 2\sigma(I)]$	0.0449
w$R2$	0.1081

$a = b = c = 10.5354(4)$ Å and $Z = 4$. The crystallographic agreement factors are $R1 = 0.0449$ $[I > 2\sigma(I)]$ and w$R2 = 0.1081$ (Further details are available from the Fachinformationszentrum Karlsruhe, D-76344 Eggenstein-Leopoldshafen: crysdata@fiz-karlsruhe.de by quoting the depository number CSD 417499). Table 1.1 shows the crystal data and the refinement details. The asymmetric unit contained manganese atom [Mn(1)] at position 4a (0, 0, 0), iron atom [Fe(1)] at position 4b (1/2, 1/2, 1/2), rubidium atom [Rb(1)] at position 8c (1/4, 1/4, 1/4), cyanonitrogen [N(1)] and cyanocarbon [C(1)] at positions 24e (0.3174(4), 0, 0) and 24e (0.2068(4), 0, 0), respectively, oxygen atoms of the ligand water molecule [O(1)] at a vacancy, and oxygen atoms of zeolitic water molecule [O(2)] at position (0.369(4), 0.369(4), 0.369(4)) and [O(3)] at position (0.327(4), 0.327(4), 0.327(4)). Figure 1.2 shows the -plane of unit cell for the cubic network. All the cyanide groups exist as bridges between Mn(1) and Fe(1) in the three-dimensional framework. Mn(1) is connected to N(1) and O(1) for an average composition of $MnN_{0.87}O_{0.13}$ due to the vacancy of $[Fe(CN)_6]^{3-}$. The interatomic distances of Fe(1)–C(1), C(1)–N(1), and Mn(1)–N(1) [or Mn(1)–O(1)] are 1.9238(1), 1.1652(1), and 2.1786(1) Å, respectively. Rb(1) occupies the center of the interstitial sites, and O(2) and O(3) are distributed in a disordered fashion inside the Mn(1)–N(1) [or Mn(1)–O(1)] interstitial sites of the cubic network as zeolitic waters.

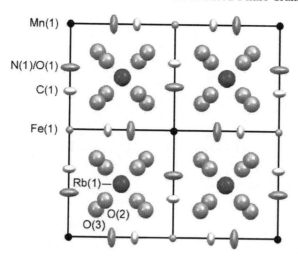

Fig. 1.2. Crystal structure for $Rb_{0.61}Mn[Fe(CN)_6]_{0.87} \cdot 1.7H_2O$. The projection in the -plane (cubic,Fm$\bar{3}$m). Spheres and ellipsoids are drawn at a 50% probability level. All H atoms are omitted for clarify. Occupancies are 0.305 for Rb(1), 1.00 for Mn(1), 0.87 for Fe(1), 0.87 for C(1), 0.87 for N(1), 0.13 for O(1), 0.0528 for O(2), and 0.0624 for O(3), respectively

1.4 Temperature-Induced Phase Transition

1.4.1 Phase Transition Phenomenon in Magnetic Susceptibility

Figure 1.3 shows the product of the molar magnetic susceptibility (χ_M) and the temperature (T) vs. T plots of RbMn[Fe(CN)$_6$]. The $\chi_M T$ value in the high-temperature (HT) phase is 4.67 cm^3 K mol^{-1} at 330 K, but cooling the sample at a cooling rate of 0.5 K min^{-1} decreases the $\chi_M T$ value around 235 K and at $T = 180$ K in the low-temperature (LT) phase reaches 3.19 cm^3 K mol^{-1}. Conversely, as the sample in the LT phase is warmed at a heating rate of 0.5 K min^{-1}, the $\chi_M T$ value suddenly increases near 285 K and reaches the HT phase value at 325 K. The transition temperatures from HT to LT ($T_{1/2\downarrow}$) and from LT to HT ($T_{1/2\uparrow}$) are 225 and 300 K, respectively, and the width of the thermal hysteresis loop ($\Delta T = T_{1/2\uparrow} - T_{1/2\downarrow}$) is 75 K. This temperature-induced phase transition is repeatedly observed [55, 56].

1.4.2 Change in Electronic State

X-ray photoelectron spectroscopy (XPS) spectra of K$^I_3$[Fe^{III}(CN)$_6$], K$^I_4$[Fe^{II}(CN)$_6$] and the HT and LT phases were measured. In the HT phase, the Fe–2P$_{3/2}$ and Mn–2P$_{3/2}$ electron binding energies are 710.1 and 641.8 eV, respectively, and in the LT phase, the Fe–2P$_{3/2}$ and Mn–2P$_{3/2}$ electron binding energies are 708.8 and 642.5 eV, respectively. The observed Fe–2P$_{3/2}$ electron binding energy of 710.1 eV in the HT phase corresponds to that of

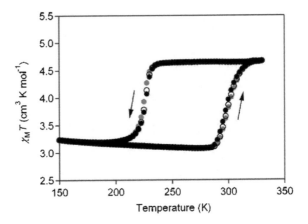

Fig. 1.3. The observed $\chi_M T - T$ plots under 5000 Oe with the first measurement (*black circles*), second measurement (*white circles*), and third measurement (*gay circles*). The *down* and *up arrows* indicate cooling and warming processes, respectively

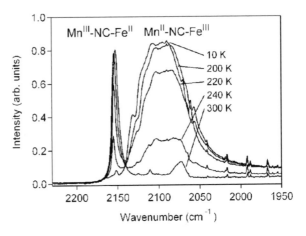

Fig. 1.4. Temperature dependence of the CN stretching frequencies in the IR spectra with cooling process

710.0 eV for Fe^{III} in $K^I_3[Fe^{III}(CN)_6]$. In contrast, $Fe–2P_{3/2}$ binding energy of 708.8 eV in the LT phase is close to that of 709.1 eV for Fe^{II} in $K^I_4[Fe^{II}(CN)_6]$. The shift of the $Mn–2P_{3/2}$ binding energy from the HT to the LT phases suggests that the oxidation number of the Mn ion increases from II to III.

Between 300 and 10 K, the infrared (IR) spectra are recorded. Figure 1.4 shows the CN^- stretching frequencies at 300, 240, 220, 200, and 10 K. At 300 K, a sharp CN^- peak is observed at 2152 cm^{-1} (linewidth= 9 cm^{-1}) and as the temperature decreases, the intensity of this peak decreases. Near 220 K a new broad peak appears at 2095 cm^{-1} (linewidth= 65 cm^{-1}). These IR changes are in the same temperature range of the phase transition in the $\chi_M T - T$ plots. The CN stretching peak at 2152 cm^{-1} in the HT phase is due

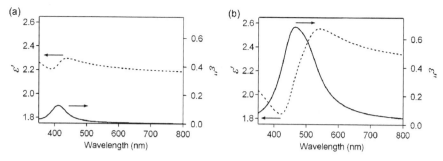

Fig. 1.5. Real (ε') and imaginary (ε'') parts of the dielectric constant (ε) spectra in the **(a)** HT phase and **(b)** LT phase. *Gray* and *black lines* represent the and parts, respectively

to the CN ligand bridged to Mn^{II} and Fe^{III} ions (Mn^{II}-NC-Fe^{III}). In contrast, the broad CN stretching peak at 2095 cm^{-1} in the LT phase is assigned to the CN ligand bridged to Mn^{III} and Fe^{II} ions (Mn^{III}-NC-Fe^{II}).

These XPS and IR spectra show that valence states for Mn and Fe ions in the HT phase are $Mn^{II}(d^5)$ and $Fe^{III}(d^5)$, respectively, and those in the LT phase are $Mn^{III}(d^4)$ and $Fe^{II}(d^6)$, respectively. The drop in the $\chi_M T$ value at $T_{1/2\downarrow}$ implies that the electronic states of the HT and LT phases are $Mn^{II}(d^5; S = 5/2)-NC-Fe^{III}(d^5; S = 1/2)$ and $Mn^{III}(d^4; S = 2)-NC-Fe^{II}(d^6; S = 0)$, respectively. These assignments are confirmed by Mn and Fe 3p-1s X-ray emission spectroscopy [57] and 1s X-ray absorption spectroscopy [58].

Figure 1.5a shows the real (ε') and imaginary (ε'') parts of the dielectric constant (ε)spectrum of the HT phase at 293 K, measured by spectroscopic ellipsometry [59]. A dispersive-shaped line, which was centered at 410 nm, was observed. In the corresponding position, an absorption-shaped peak was observed in the ε'' spectrum at 410 nm with $\varepsilon'' = 0.13$. This peak is assigned to the ligand-to-metal charge transfer (LMCT) transition of $[Fe(CN)_6]^{3-}$ ($^2T_{2g} \rightarrow^2 T_{1u}, CN^- \rightarrow Fe^{III}$). Figure 1.5b shows the ε' and ε'' parts of the ε spectrum in the LT phase. The LT phase was obtained by slowly cooling to 160 K using N_2 vapor, and then measuring ε at 275 K. A large dispersive-shaped line, which was centered at 470 nm with a minimum value at 420 nm and a maximum at 540 nm, was observed in the ε' spectrum. The corresponding position in the ε'' spectrum showed a strong absorption-shaped peak of $\varepsilon'' = 0.68$, which is assigned to the metal-to-metal charge transfer (MM'CT) band of $Fe^{II} \rightarrow Mn^{III}$ (more accurately, $CN_{2px}, CN_{2py} \rightarrow Mn_{3dx^2-y^2}, Mn_{3dz^2}$).

1.4.3 Structural Change

Figure 1.6 shows the powder X-ray diffraction (XRD) patterns as the temperature decreased from 300, 240, 220 to 160 K. The diffraction pattern of the HT phase is consistent with a face-centered cubic (F$\bar{4}$3m) structure with a lattice constant of 10.533 Å(at 300 K). As the sample is cooled, the XRD

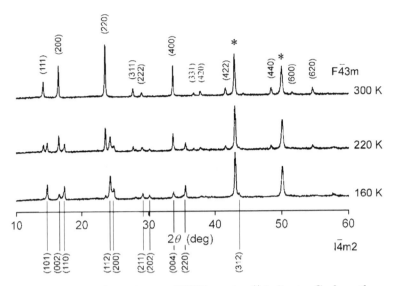

Fig. 1.6. Temperature dependences of XRD spectra (* indicates Cu from the sample holder)

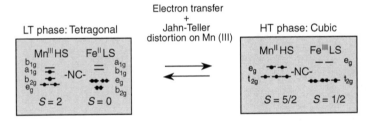

Fig. 1.7. Electronic states of the LT and HT phases

peaks of the HT phase decrease and different peaks appear. The observed XRD pattern in the LT phase shows a tetragonal structure of I4̄m2 with $a = b = 7.090$ Å and $c = 10.520$ Å (at 160 K), which corresponds to $a = b = 10.026$ Å and $c = 10.520$ Å in a cubic lattice. The unit cell volume of 1169 Å^3 in the HT phase is reduced about 10% to 11,057 Å^3 in the LT phase and warming caused the tetragonal structure to return to the cubic one. This structural change from cubic to tetragonal in the XRD measurement is understood by the Mn^{III} Jahn–Teller transformation of the tetragonally octahedral elongation-type (B$_{1g}$ oscillator mode). Synchrotron radiation X-ray powder structural analysis was used to determine the precise bond lengths of the LT phase, i.e., two-long and four-short Mn–N bond distances are 2.26(2) and 1.89(3) Å, respectively, and the two-short and four-long Fe–C bond distances are 1.89(2) and 2.00(3), respectively [60]. Thus, the d-orbital symmetry of both metal ions in the LT phase is D_{4th} (a$_{1g}$,b$_{1g}$,b$_{2g}$,and eg). Therefore, the precise electronic state of LT phase is Mn^{III}(e$_g$^{2}b$_{2g}$^{1}a$_{1g}$^{1};$S = 2$)–NC–Fe^{II}(b$_{2g}$^{2}e$_g$^{4};$S=0$) (Fig. 1.7).

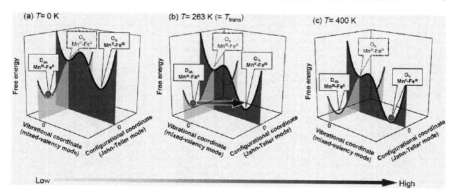

Fig. 1.8. The schematic free energy surfaces of this system in mixed-valence (*black curve*) and Jahn–Teller (*gray curve*) modes: (**a**) the ground state is $Mn^{III} - Fe^{II}$ and the meta-stable state is $Mn^{II} - Fe^{III}$ at $T = 0$ K, (**b**) $T = 263$ K ($=T_{trans}$), (**c**) the ground state is $Mn^{II} - Fe^{III}$ and meta-stable state is $Mn^{III} - Fe^{II}$ at $T = 400$ K. Gray spheres indicate population

1.4.4 Mechanism

Prussian blue analogs belong to class II mixed-valence compounds. This system is described by two parabolic potential-energy curves due to valence isomers in the nuclear coordinates of the coupled vibrational mode [8–11]. When these two vibronic states interact, the ground state surface has two minima in the vibrational coordinates (Fig. 1.8). In the present system, the $Mn^{III} - Fe^{II}$ vibronic state is a ground state at $T = 0$ K in the vibrational coordinates (mixed-valency mode) (black curve in Fig. 1.8a). Moreover, in this situation, Mn^{III} causes Jahn–Teller distortion, and then the energy of the $Mn^{III} - Fe^{II}$ has two minima described in the configurational coordinates (Jahn–Teller mode) (gray curve in Fig. 1.8a). In the present system, Mn^{III} ion shows an enlongation-type Jahn–Teller distortion. These potential surfaces change as the temperature increases, which cause a phase transition.

1.5 Ferromagnetism of the Low-Temperature Phase

1.5.1 Magnetic Ordering and Heat Capacity

When the LT phase is cooled to a very low temperature under an external magnetic field of 10 Oe, it exhibits spontaneous magnetization with a Curie temperature (T_c) of 11.3 K (Fig. 1.9a). The magnetization as a function of the external magnetic field at 3 K indicates that the saturated magnetization (M_s) value is 3.6 μ_B and the coercive field (H_c) value is 1050 G (Fig. 1.9b). The $\chi_M^{-1} - T$ plots of the paramagnetic LT phase show positive Weiss temperatures (Θ) between 12 and 15 K, which are obtained by extrapolating the

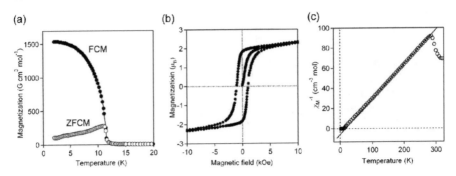

Fig. 1.9. (a) Magnetization vs. temperature plots of the LT phase by SQUID measurement: (•) Field-cooled magnetization (FCM) at 10 Oe, and (○) zero field-cooled magnetization (ZFCM) at 10 Oe. (b) Magnetic hysteresis loop of the LT phase at 3 K. (c) The observed $\chi_M - T$ plots. The data between 150 and 270 K is fitted to Curie-Weiss plots (−)

data in the temperature region of 150–270 K, respectively [61]. In low temperature region, the C_p value gradually increases with temperature and reaches a maximum, 27.1 J K^{-1} mol^{-1} at 11.0 K (denoted here as T_p), as shown in Fig. 1.10a. Then it drops suddenly to 17.5 J K^{-1} mol^{-1}, and increases gradually. The dependence of the C_p values on the external magnetic field is shown in Fig. 1.10b and c, where the T_p peaks shift to a higher temperature as the external magnetic field increases: $T_p = 11.0$ K (= 0 T), 11.0 K (0.05 T), 11.2 K (0.10 T), 11.3 K (0.20 T), 11.4 K (0.30 T), 11.5 K (0.50 T), 11.9 K (1.00 T), 13.8 K (2.00 T), and 15.2 K (3.00 T).

1.5.2 Entropy and Enthalpy of Magnetic Phase Transition

Because RbMn[Fe(CN)$_6$] is an insulating magnetic system, the C_p value is described as a sum of the contributions from lattice vibration, C_{lat}, short-range magnetic ordering, C_{short}, and long-range magnetic ordering, C_{long} :

$$C_p = C_{lat} + C_{short} + C_{long}. \tag{1.1}$$

C_{lat} is described by a polynomial function of temperature with odd powers [62],

$$C_{lat} = aT^3 + bT^5 + cT^7 + dT^9 + eT^{11} + \cdots, \tag{1.2}$$

and C_{short} is described by AT^{-2} [63]. We fitted the C_p data in the region between 15 K (= 1.4 × Tc) and 30 K (= 2.7 × Tc) by the contributions of C_{lat}+Cshort, using analyses reported in other systems [64]. The derived coefficients, including the estimated uncertainties (±7.4%) from the experiment (±7.0%) and curve fitting (±2.3%), are as follows: $a = 8.08 \times 10^{-3}$ J K^{-4} mol−1, $b = -2.10 \times 10^{-5}$ J K^{-6} mol^{-1}, $c = 2.56 \times 10^{-8}$ J K^{-8} mol^{-1}, $d = -1.18 \times 10^{-11}$ J K^{-10} mol^{-1}, and $A = 1130$ J K mol^{-1}. The

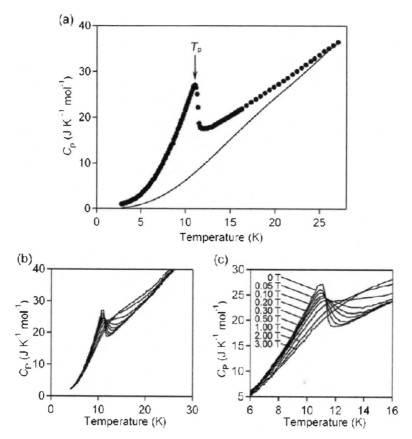

Fig. 1.10. (a) Plots of C_p vs. T in a zero external magnetic field: (●) experimental and (−) derived C_{lat} curve based on (1.3). (b) Plots of C_p vs. T in the presence of an external magnetic field. (c) Enlarged plots of (b)

solid line in Fig. 1.10a shows the C_{lat} curve. The magnetic heat capacity , $C_{mag} = C_{short} + C_{long}$, is obtained by subtracting C_{lat} from C_p, as shown in Fig. 1.11. The magnetic transition entropy, ΔS_{mag}, and enthalpy, ΔH_{mag}, can be obtained from

$$\Delta S_{mag} = \int_0^T C_{mag} d\ln T \tag{1.3}$$

and

$$\Delta H_{mag} = \int_0^T C_{mag} dT. \tag{1.4}$$

The estimated values of ΔS_{mag} and ΔH_{mag} for $Rb^I Mn^{III}[Fe^{II}(CN)_6]$ are 11.8 ± 0.9 J K^{-1}mol^{-1} and 125 ± 9 J mol^{-1}, respectively.

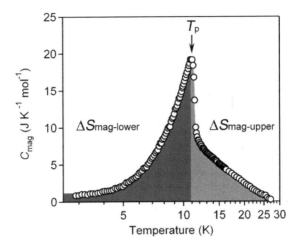

Fig. 1.11. Plots of C_{mag} vs. log T

1.5.3 Long-Range Magnetic Ordering and Exchange Coupling

The T_p value of 11.0 K agrees with the T_C value of 11.3 K derived from a SQUID measurement, then, the anomalous peak at can be ascribed to a magnetic phase transition. The ΔS_{mag} value of 11.8±0.9 J K^{-1}mol^{-1} is close to the value calculated for the ordering of magnetic spins on the Mn^{III}(S = 2) sites for Rb^{I}Mn^{III}[Fe^{II}(CN$_6$)] given by $R\ln(2S+1)$=13.4 J K^{-1}mol^{-1}, where R is the gas constant. Thus, the origin of this magnetic phase transition is attributed to the long-range magnetic ordering of the Mn^{III} sites.

The dimensionality of magnetic ordering, i.e., two- or three-dimensional (2- or 3-D) magnetic lattice, can be determined by the temperature dispersion of ΔS_{mag}. When the value of ΔS_{mag} is divided into two terms, such as the magnetic entropy values below T_p ($\Delta S_{mag-lower}$) and above T_p ($\Delta S_{mag-upper}$), the ratio of $\Delta S_{mag-lower}$ / ΔS_{mag} for the magnetic lattices of the 3-D Ising, 2-D Ising, and 3-D Heisenberg types are 81, 44, and 62%, respectively [65]. The ratio of $\Delta S_{mag-lower}$ / ΔS_{mag} in the present system is 65(3)% (Fig. 1.11). Therefore, in this framework the magnetic ordering of the LT phase is most likely 3-D Heisenberg-type magnetic ordering.

To analyzing C_{mag} at very low temperatures using the spin-wave theory can determine if the long-range magnetic ordering of a target material is ferromagnetic or antiferromagnetic. The heat capacity due to the spin-wave excitation,C_{SW}, is expressed by [66]:

$$C_{SW} = \alpha T^{\frac{d}{n}}, \qquad (1.5)$$

where d stands for the dimensionality of the magnetic lattice and n is the exponent in the dispersion relationship: $n = 1$ for antiferromagnets and $n =$

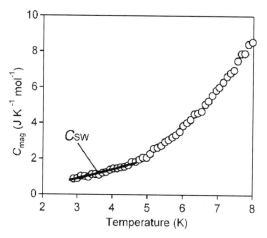

Fig. 1.12. Experimental plots of C_{mag} ($\circ$) and the C_{SW} curve (–) calculated from the spin-wave theory for a 3-D ferromagnet using (1.6) with $d/n = 1.51$ and $\alpha = 0.17\text{J K}^{-5/2}\text{mol}^{-1}$

2 for ferromagnets. We fitted the C_{mag} values in the region between 2.8 and 4.7 K to (1.5) (Fig. 1.12) and the estimated parameter of d/n is 1.51(11). This d/n value is consistent with that predicted for the magnetic ordering of the LT phase, i.e., the 3-D ferromagnet, where $d = 3$ and $d = 2$. The observed shifts in the T_p values of the in-field C_p data, 11.0 K (0 T) $\rightarrow$ 15.2 K (3.00 T) displayed in Fig. 1.10b and c, which also suggest ferromagnetic character. Since the shift in T_p to higher temperatures is characteristic of ferromagnetic transitions [67], the trend of the in-field C_p values observed in the present study gives direct evidence that the magnetic ordering of the LT phase is ferromagnetic.

This system shows a 3-D Heisenberg-type ferromagnetic ordering, although diamagnetic Fe^{II} is bridged to Mn^{III} in an alternating fashion. The exchange coupling constant, J, of this ferromagnet can be evaluated in the following manner. The α value derived from (1.5) is related to the J value. In C_{SW} for a 3-D ferromagnet, the coefficient α is described by [68]:

$$\alpha = \frac{1}{\sqrt{2}} \frac{5R\zeta(5/2)\Gamma(5/2)}{16\pi^2 S^{3/2}} \left(\frac{k_B}{J}\right)^{3/2}, \tag{1.6}$$

where ζ is Riemann's zeta function, Γ is Euler's gamma function, and k_B is the Boltzmann constant. Since the α value obtained from (1.5) is 0.17(1)J K$^{-5/2}$ mol^{-1}, the estimated J value based on (1.6) is +0.55(4) cm^{-1}. ΔH_{mag} is also related to the J value in an extension of the molecular-field theory. In this treatment, ΔH_{mag} due to long-range magnetic ordering is expressed by

$$\frac{\Delta H_{mag}}{R} = \frac{S^2 zJ}{k_B}, \tag{1.7}$$

where the number of neighboring magnetic sites, z, is 6 in the present system. The estimated J value from (1.7), using $\Delta H_{mag} = 125 \pm 9$ Jmol^{-1} is $+0.44(3)$ cm^{-1}.

1.5.4 Mechanism of Magnetic Ordering

Application of the superexchange interaction mechanism to the present ferromagnetic ordering is difficult since the diamagnetic Fe^{II} sites are connected by paramagnetic Mn^{III} sites. One plausible mechanism is the valence delocalization mechanism, in which ferromagnetic coupling arises from the charge-transfer configuration [69]. Day et al. explained the ferromagnetism of Fe^{III}[Fe^{II}(CN)6]$_{0.75}$ · 3.5H$_2$O by the ferromagnetic exchange interaction based on a partial delocalization of the electrons that occupy the Fe^{II}t$_{2g}$ orbitals next to the neighboring high-spin Fe^{III} sites. Since Fe^{III} in Prussian blue is replaced with Mn^{III}, the same mechanism is feasible in our system. In fact, an intense intervalence transfer (IT) band of the LT phase has been observed at 540 nm and in the IT band of Prussian blue. In the valence delocalization mechanism, the T_c value is related to the valence delocalization coefficient of c as $T_c \propto c^4$. The c value is given by second-order perturbation theory as

$$c = \sum_{i=2,3} \left(\langle \psi_0 | H | \psi_i \rangle \langle \psi_1 | H | \psi_i \rangle / (E_1 - E_0)(E_i - E_0) \right), \qquad (1.8)$$

where ψ_0, ψ_1, ψ_2, and ψ_3 are the ground (pure Mn^{III} − Fe^{II}) state and the charge-transfer configurations of Fe^{II} → Mn^{III}, Fe^{II} → CN, and CN → Mn^{III}, respectively, and $E_0 - E_3$ are their energies. Mixing these excited charge-transfer configurations with the ground state causes the ferromagnetic exchange coupling. The J value of $\approx +0.5$ cm^{-1} in the present system is three times larger than that of $+1.5$ cm^{-1} in Prussian blue. This large J value means that Rb^{I}Mn^{III}[Fe^{II}(CN)6] has a large c value. Namely, the electrons on the Fe^{II} site are delocalized to the Mn^{III} site.

1.6 Control of Temperature-Induced Phase Transition

1.6.1 Huge Thermal Hysteresis Loop and a Hidden Stable Phase

Rb$_{0.64}$Mn[Fe(CN)6]$_{0.88}$ · 1.7H$_2$Owas prepared by reacting an aqueous solution (0.1 mol dm^{-3}) of MnCl$_2$ with a mixed aqueous solution of RbCl (1.0 mol dm^{-3}) and K$_3$[Fe(CN)6] (0.1 mol dm^{-3}). The mixed solution was stirred for 5 min and the precipitate was filtered, yielding a light brown powder. The SEM image indicates that the precipitate is rectangular with the size of 0.3 ± 0.1 μm. The IR peak is observed at 2153 cm^{-1} at 300 K, which is assigned to the CN group of Fe^{III} − NC − Mn^{II}, i.e., the electronic state of the prepared compound is Rb$^{I}_{0.64}$Mn^{II}[Fe^{II}(CN)6]$_{0.88}$ · 1.7H$_2$O.

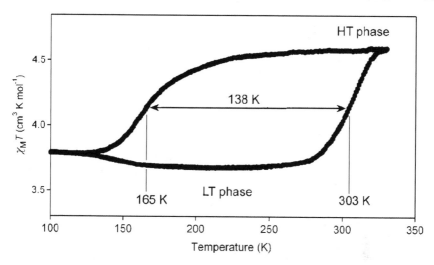

Fig. 1.13. The observed $\chi_M T$ vs. T plots for $Rb_{0.64}Mn[Fe(CN)_6]_{0.88} \cdot 1.7H_2O$ under 5000 Oe with cooling and warming by $\pm$ 0.5 K min^{-1}

The magnetic properties were measured using a SQUID magnetometer. Figure 13 shows the product of the $\chi_M T$ vs. T plots. The $\chi_M T$ value decreases around 165 K ($= T_{1/2\downarrow}$) as the sample is cooled at a cooling rate of –0.5 K min^{-1}. Conversely, as the sample is warmed at a warming rate of +0.5 K min^{-1}, the $\chi_M T$ value increases around 303 K ($= T_{1/2\uparrow}$) and returns to the initial value. The thermal hysteresis value ($\Delta T \equiv T_{1/2\uparrow} - T_{1/2\downarrow}$) is surprisingly large, 138 K. In addition, the χ_M value of the rapidly cooled sample, i.e., the sample placed directly into a sample chamber at 10 K, was measured. The rapidly cooled sample shows a high $\chi_M T$ value even at low temperature (hereafter called the hidden stable phase), which nearly corresponds to the value extrapolated from the HT phase, and then relaxes to the $\chi_M T$ value of low-temperature (LT) phase around 114 K ($= T_{SP\downarrow}$) (Fig. 1.14a). Since the χ_M^{-1} vs. T plots of the HT and LT phases are nearly linear as a function of T, these plots are fitted by the Curie-Weiss law and the Weiss temperatures of the HT and LT phases, which are estimated to be –6 K and +7 K, respectively (Fig. 1.14b) [70].

To investigate the electronic state of the LT phase, the temperature dependence of the CN stretching frequencies in the IR spectrum was measured. As the temperature decreases, the intensity of the $Mn^{II} - NC - Fe^{III}$ peak at 2153 cm^{-1} decreases and a new broad peak appears between 2080 and 2140 (peak top: 2108 cm^{-1}), which is assigned to the CN group of $Mn^{III} - NC - Fe^{II}$. From the analysis of the conversion in the IR spectra, the electronic state of the LT phase is determined to be $Rb^{I}_{0.64}Mn^{II}_{0.40}Mn^{III}_{0.60}[Fe^{II}(CN)_6]_{0.60}[Fe^{III}(CN)_6]_{0.28} \cdot 1.7H_2O$. The XRD

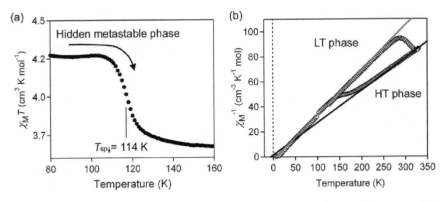

Fig. 1.14. (a) The observed $\chi_M T$ vs. T plots for $Rb_{0.64}Mn[Fe(CN)_6]_{0.88} \cdot 1.7H_2O$ under 5000 Oe with warming by $+0.1$ K min^{-1} after rapid-cooling. **(b)** The observed χ_M^{-1} vs. T plots, and the χ_M^{-1} vs. T curves of the HT (*black*) and the LT (*gray*) phases, fitted by Curie-Weiss law

pattern of the sample at 300 K shows a cubic crystal ($Fm\bar{3}m$) with a lattice constant of $a = 10.535(6)$ Å. As the temperature decreases, different XRD patterns due to the LT phase appear near $T_{1/2\downarrow}$. The observed XRD pattern in the LT phase is assigned to an orthorhombic crystal structure (F222) with lattice constants of $a = 10.261(16)$, $b = 10.044(10)$, and $c = 10.567(16)$ Å. This distorted crystal structure is ascribed to the Jahn–Teller effect on the produced Mn^{III} sites.

1.6.2 Thermodynamical Analysis of Thermal Hysteresis Loop

The fractions (α) of the temperature-induced phase transition of $\Delta T = 138$ K are estimated as shown in Fig. 1.15a using the extrapolation curves of vs. χ_M^{-1} plots of the HT and LT phases. As a reference, χ_M^{-1} vs. T of $RbMn[Fe(CN)_6]$ ($T_{1/2\downarrow} = 231$ K, $T_{1/2\uparrow} = 304$ K, and $\Delta T = 73$ K) from our previous work [55] is also s hown in Fig. 1.15b. These thermal hysteresis loops are analyzed based on SD model [71]. The Gibbs free energy G of the system is described by $G = \alpha\Delta H + \gamma\alpha(1-\alpha) + T\{R[\alpha\ln\alpha + (1-\alpha)\ln(1-\alpha)] - \alpha\Delta S\}$, taking G of the LT phase as the origin of the energies, where α is the fraction of the HT phase, ΔH is the transition enthalpy, ΔS is the transition entropy, R is the gas constant, and the γ is an interaction parameter as a function of temperature, $\gamma = \gamma_a + \gamma_b T$ [71, 72]. Experimental heat capacity measurements indicate that the ΔH and ΔS values of $RbMn[Fe(CN)_6]$ are $\Delta H = 1.7$ KJ mol^{-1} and $\Delta S = 6.0$ J K^{-1}, respectively [55, 56]. When these thermodynamic parameters are used, the thermal hysteresis loops of $Rb_{0.64}Mn[Fe(CN)_6]_{0.88} \cdot 1.7H_2O$ and $RbMn[Fe(CN)_6]$ are well reproduced with the parameters of $(\Delta H, \Delta S, \gamma_a, \gamma_b) = (1.24$ kJ mol^{-1}, 4.54 J K^{-1} mol^{-1}, 20.1 kJ mol^{-1}, 12.0 J K^{-1} mol^{-1}) and $(\Delta H, \Delta S, \gamma_a, \gamma_b) = (1.68$ kJ

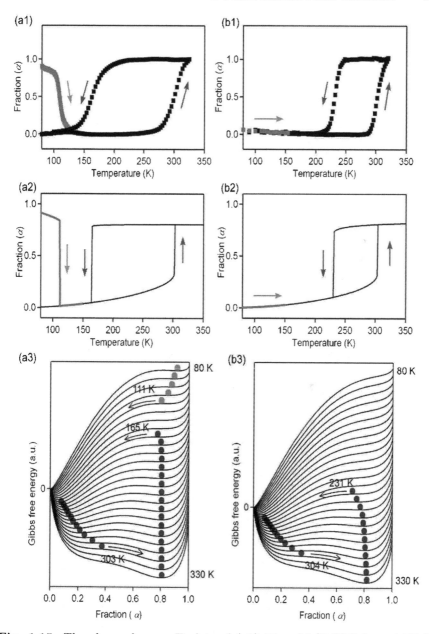

Fig. 1.15. The observed α vs. T plots of (**a1**) $Rb_{0.64}Mn[Fe(CN)_6]_{0.88} \cdot 1.7H_2O$ and (**b1**) $RbMn[Fe(CN)_6]$ from [55] where the *gray circles* show the rapidly-cooled sample upon warming. Calculated thermal hysteresis loops of (**a2**) $Rb_{0.64}Mn[Fe(CN)_6]_{0.88} \cdot 1.7H_2O$ and (**b2**) $RbMn[Fe(CN)_6]$. Temperature dependence of calculated vs. curves between 80 and 330 K with 10 K interval for (**a3**) $Rb_{0.64}Mn[Fe(CN)_6]_{0.88} \cdot 1.7H_2O$ and (**b3**) $RbMn[Fe(CN)_6]$. The *circles* indicate the thermal populations. The *black* and *gray circles* indicate the temperature-induced phase transition and the relaxation, respectively

mol^{-1}, 6.0 J K^{-1} mol^{-1}, 20.5 kJ mol^{-1}, 11.9 J K^{-1} mol^{-1}), respectively (Figs. 1.15(a2),(a3), black lines). In addition, Rb$_{0.64}$Mn[Fe(CN)$_6$]$_{0.88}$ · 1.7H$_2$O shows that a hidden stable phase exists at low temperature under thermal equilibrium condition (Fig. 1.15(a2), gray line). In this low temperature region, a local energy minimum exists at α = 0.85 – 0.9 and relaxes to LT phase at 111 K (Fig. 1.15(a3), gray circles), which well reproduces the experimental data (Fig. 1.15(a1), gray circles). In contrast, both the calculated and experimental data of RbMn[Fe(CN)$_6$] indicate that this hidden stable phase does not exist (Fig. 1.15b). These results suggest that the observed phase in Rb$_{0.64}$Mn[Fe(CN)$_6$]$_{0.88}$ · 1.7H$_2$O is a hidden stable state of HT phase under thermal equilibrium condition and is not a supercooled phase under nonequilibrium condition. Furthermore, we calculated the α vs. T plots with various parameters, and thus concluded that only the system showing a large thermal hysteresis loop produces the hidden stable phase under thermal equilibrium condition.

1.7 Photo-Induced Phase Collapse

1.7.1 Non Phase Transition Material

Rb$_{0.43}$Mn[Fe(CN)$_6$]$_{0.81}$ · 3H$_2$O was prepared by reacting an aqueous solution (0.1 mol dm^{-3}) of MnCl2 with a mixed aqueous solution of RbCl (0.05 mol dm^{-3}) and K$_3$[Fe(CN)$_6$] (0.1 mol dm^{-3}). The temperature dependence of the magnetic susceptibility of the initial Mn^{II} – Fe^{III} phase was measured using a SQUID magnetometer, and Fig. 1.16 plots the product of the $\chi_M T$ vs. T at a very slow cooling rate of –0.05 K min^{-1}. In the $\chi_M T - T$ plots, $\chi_M T$ remained nearly constant, corresponding to the sum of Mn^{II}(S = 5/2) and Fe^{III}(S = 1/2). Variable temperature IR spectra also showed that the CN streching frequency peak due to Mn^{II} – NC – Fe^{III} is maintained down to low temperature. The

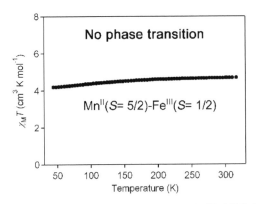

Fig. 1.16. The observed $\chi_M T$ vs. T plots for Rb$_{0.43}$Mn[Fe(CN)$_6$]$_{0.81}$ · 3H$_2$O under 5000 Oe

XRD pattern at 300 K confirmed that the crystal structure is face-centered cubic (space group: Fm3̄m) with a lattice constant of $a = 10.473(9)$Å. In the temperature range between 300 and 20 K, the lattice constant was almost constant, i.e., $a = 10.493(9)$ Å at 20 K. These results indicate that a temperature-induced charge-transfer phase transition does not occur in the present material [73]. In spectroscopic ellipsometry , an optical resonance due to the ligand-to-metal charge transfer (LMCT) transition on $[Fe(CN)_6]^{3-}$ was observed at 410 nm.

1.7.2 Photo-Induced Structural Transition

Since $Rb^I_{0.43}Mn^{II}[Fe^{III}(CN)_6]_{0.81} \cdot 3H_2O$ has absorption at 410 nm, we irradiated the sample in XRD equipment with blue light (410±25 nm, 20 mW cm^{-2}) using a filtered Xe lamp. Upon blue-light irradiation, the XRD peaks of the $Mn^{II} - Fe^{III}$ phase decreased, and new XRD peaks appeared as shown in Fig. 1.17. The new XRD pattern of the PG phase was assigned to a face-centered cubic structure of Fm3̄m with $a = 10.099(3)$Å. When the XRD pattern due to the PG phase was cooled to 20 K and then warmed above room temperature, it was maintained over a wide temperature range, but at 310 K the XRD pattern was perfectly restored to the original XRD pattern of the initial $Mn^{II} - Fe^{III}$ phase (Fig. 1.18a). The photo-conversion efficiency depended on the irradiation temperature as shown in Fig. 1.18b, i.e., 0% (20 K), 45% (100 K), 66% (140 K), 3% (180 K), 6% (220 K), and 0% (300 K) by blue light. To investigate the electronic state of the PG phase, the IR spectra after light

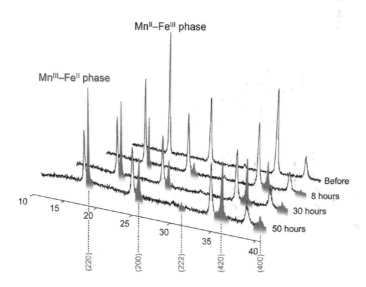

Fig. 1.17. Photo-induced phase collapse in $Rb_{0.43}Mn[Fe(CN)_6]_{0.81} \cdot 3H_2O$ by blue-light irradiation. XRD patterns at 140 K before and after blue-light irradiation

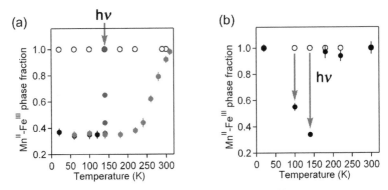

Fig. 1.18. (**a**) Temperature dependence of the $Mn^{II} - Fe^{III}$ phase fraction before ir-radiation (*open circles*), during light irradiation at 140 K (*dark gray circles*), cooling process after irradiation (140 K→20 K) (*black circles*), and warming process (20 K → 310 K) (*gray circles*). (**b**) Irradiation temperature dependence of the $Mn^{II} - Fe^{III}$ phase fraction before (*open circles*) and after (*black circles*) light irradiation for 50 h (20 mW cm^{-2})

irradiation were measured at 140 K. Upon irradiation, the $Mn^{II} - NC - Fe^{III}$ peak at 2153 cm^{-1} decreased and a broad peak appeared at 2095 cm^{-1}, which corresponds to the CN streching frequency of $Mn^{III} - NC - Fe^{II}$. Hence, the observed photo-induced phase collapse is caused by the charge-transfer phase transition from the $Mn^{II} - Fe^{III}$ phase to the $Mn^{III} - Fe^{II}$ phase.

1.7.3 Photo-Induced Phase Transition from a Metastable Phase to a Hidden Stable Phase

To understand the mechanism of the observed photo-induced phase tran-sition, we calculated the Gibbs free energy vs. the $Mn^{II} - Fe^{III}$ fraction for $Rb_{0.43}Mn[Fe(CN)_6]_{0.81} \cdot 3H_2O$ using the Slichter and Drickamer mean-field model [71], described by $G = \alpha\Delta H + \gamma\alpha(1 - \alpha) + T\{R[\alpha\ln\alpha + (1 - \alpha)\ln(1 - \alpha)] - \alpha\Delta S\}$, mentioned in Sect. 1.6.2. The thermodynamical pa-rameters for the calculation were estimated by extrapolating our previous data. In the entire temperature range, a free-energy barrier existed between the mainly $Mn^{II} - Fe^{III}$ phase and the mainly $Mn^{III} - Fe^{II}$ phase, as shown in Fig. 1.19a. Since the material synthesis was carried out at room tem-perature and produced the $Mn^{II} - Fe^{III}$ phase, the $Mn^{II} - Fe^{III}$ phase is ex-pected to be maintained in the entire temperature range as shown in Fig. 1.19a (dark gray circles). This calculation well explains the observed tem-perautre dependence in $Rb_{0.43}Mn[Fe(CN)_6]_{0.81} \cdot 3H_2O$, which does not ex-hibit a thermal phase transition. At the same time, this calculation predicts the existence of a hidden stable phase, the $Mn^{III} - Fe^{II}$ phase (Fig. 1.19a, light gray circles). Its calculated temeperature dependence (Fig. 1.19b, lower, light gray curve) corresponds well to the observed temperature dependence in Fig. 1.18a.

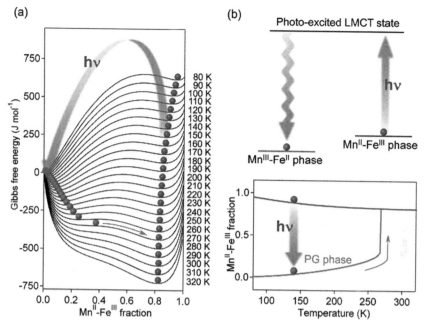

Fig. 1.19. Mechanism of the photo-induced phase collapse in $Rb_{0.43}Mn[Fe(CN)_6]_{0.81} \cdot 3H_2O$. (a) Temperature dependence of calculated Gibbs free energy vs. the $Mn^{II} - Fe^{III}$ fraction for $Rb_{0.43}Mn[Fe(CN)_6]_{0.81} \cdot 3H_2O$ based on the Slichter and Drickamer model. *Dark* and *light gray circles* indicate the existing populations of $Mn^{II} - Fe^{III}$ phase and $Mn^{III} - Fe^{II}$ phase, respectively. (b) Schematic picture of the pathway in the photo-induced phase collapse (*upper*). Temperature dependence of calculated fractions of $Mn^{II} - Fe^{III}$ mainly phase (*dark gray curve*) and $Mn^{III} - Fe^{II}$ mainly phase (*light gray curve*) (*lower*)

We thus conclude that the present photo-induced phase collapse is caused by a phase transition from a thermodynamically metastable $Mn^{II} - Fe^{III}$ phase to a $Mn^{III} - Fe^{II}$ true stable phase though the excited state of LMCT ($CN^- \rightarrow Fe^{III}$), which is excited by blue-light irradiation (Fig. 1.19b, upper).

1.8 Photo-Induced Phase Transition at Room Temperature

The photo-induced effect in the paramagnetic region was investigated with $Rb_{0.97}Mn[Fe(CN)_6]_{0.99} \cdot 0.2H_2O$ using IR spectroscopy [74, 75]. A pulsed Nd^{3+} ; YAG laser ($\lambda = 532\,nm$; pulse width: 6 ns) was used. The $\chi_M T - T$ plots showed that $Rb_{0.97}Mn[Fe(CN)_6]_{0.99} \cdot 0.2H_2O$ exhibited a temperature-induced phase transition (Fig. 1.20). The $T_{1/2\downarrow}$ and $T_{1/2\uparrow}$ were 220 and 314 K, respectively.

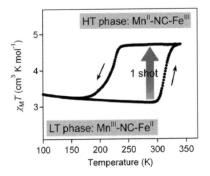

Fig. 1.20. The observed $\chi_M T - T$ plots for $Rb_{0.97}Mn[Fe(CN)_6]_{0.99} \cdot 0.2H_2O$ in the cooling ($\downarrow$) and warming ($\uparrow$) processes under 5000 Oe

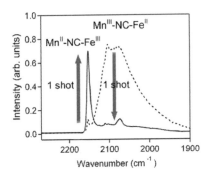

Fig. 1.21. Change in the IR spectrum by a one-shot-laser-pulse irradiation of 532 nm with 80 mJ cm^{-2} pulse^{-1} at 295 K for $Rb_{0.97}Mn[Fe(CN)_6]_{0.99} \cdot 0.2H_2O$. The spectra before and after irradiation are shown as *dotted* and *solid lines*, respectively

Figure 1.21 shows the changes in the IR spectra before and after a one-shot-laser-pulse irradiation at 295 K, which is a temperature inside the thermal hysteresis loop. A one-shot-laser-pulse irradiation changed the IR spectrum of the LT phase to that of the HT phase. The IR spectrum of the irradiated sample returned to that of the initial LT phase by cooling (295 K → 77 K → 295 K). Figure 1.22a shows the conversion fraction as a function of laser power density at 295, 280, 260, 240 and 220 K. The conversion fraction depended on the P value and temperature. A threshold in the laser power density (P_{th}) was observed. At 295 K, when the P value was greater than 6 mJ cm^{-2} pulse^{-1}, the LT phase was converted to HT phase. In contrast, in the case of $P < P_{th}$, photo-conversion did not occur even after irradiating more than thousand shots. The existence of a threshold suggests that cooperative effects exist in the present photo-induced phase transition and the maximum value of quantum yield was $\Phi = 38$ at $P = 24$ mJ cm^{-2} pulse^{-1} (Fig. 1.22b).

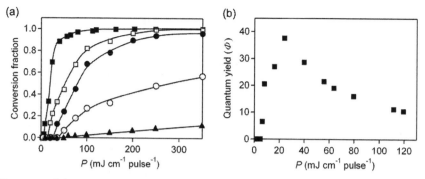

Fig. 1.22. (a) Laser power density (P) dependence of a one-shot-laser-pulse induced phase transition for $Rb_{0.97}Mn[Fe(CN)_6]_{0.99} \cdot 0.2H_2O$ when irradiating with 532 nm at 295 (■), 280 (□), 260 (●), 240 (○) and 220 K (▲). *Solid line* is for the eye guide. (b) Laser power density (P) dependence of quantum yield Φ of a one-shot-laser-pulse induced phase transition at 295 K

In rubidium manganese hexcyanoferrate, temperature-induced phase transition is observed with large thermal hysteresis loops. The charge transfer from Mn^{II} to Fe^{III} accompanying the Jahn–Teller effect on the $Mn^{III}N_6$ moiety explains this phase transition. In paramagnetic state of this system, the photo-induced phase transition inside the thermal hysteresis loop was observed at room temperature. Such a photo-induced phase transition phenomena are caused by: (1) the change of valence states on transition metal ions due to a metal-to-metal charge-transfer and (2) the bistability due to the Jahn–Teller distortion of Mn^{III} ion.

1.9 Photomagnetism

1.9.1 Photo-Induced Demagnetization by One-Shot-Laser-Pulse

In this section, the photomagnetic effect of the LT phase was investigated with $Rb_{0.88}Mn[Fe(CN)_6]_{0.96} \cdot 0.6H_2O$ using SQUID magnetometer [38, 76]. A pulsed Nd^{3+} ; YAG laser ($\lambda = 532$ nm; pulse width: 6 ns) was guided by optical fiber into the SQUID magnetometer. As $Rb_{0.88}Mn[Fe(CN)_6]_{0.96} \cdot 0.6H_2O$ was cooled to a very low temperature under an external magnetic field of 10 Oe, the LT phase exhibited spontaneous magnetization with a T_c of 12 K. The Ms and Hc values at 2 K were 3.6 μ_B and 1800 G, respectively. This Ms value can be explained by the ferromagnetic spin ordering of Mn^{III} ($S = 2$) ions. From the χ_M^{-1} vs. T plots at temperature between 100 and 250 K, the positive Weiss temperature value of +15 K was obtained.

When the sample was irradiated by one-shot-laser-pulse with 532 nm-laser light at 3 K, the magnetization was decreased. Figure 1.23a shows the magnetization vs. temperature curve for the sample irradiated with $P = 130$ mJ cm^{-2}

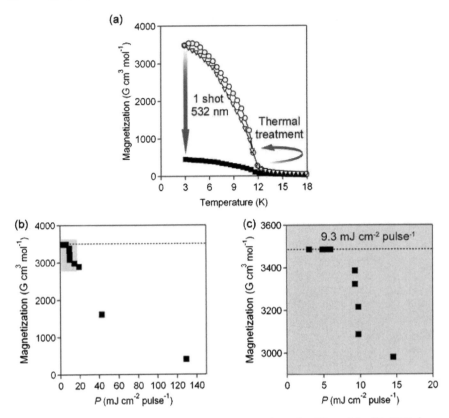

Fig. 1.23. (a) Magnetization vs. temperature plots for Rb$_{0.88}$Mn [Fe(CN)$_6$]$_{0.96}$·
0.6H$_2$O at 200 Oe before (○) and after the one-shot-laser-pulse irradiation (■) and
thermal treatment (▽). (b) Laser power density (P) dependences of the one-shot-
laser-pulse induced photodemagnetization phenomenon. (c) Enlarged plots of (b)

pulse $^{-1}$. The photo-conversion increased with increasing the laser power density (P) as shown in Fig. 1.23b and c. A threshold in the laser power density
(P'_{th}) was observed; when the P value was above 9.3 mJ cm^{-2} pulse $^{-1}$, the
magnetization value was decreased, however, in the case of $P < P'_{th}$, the
magnetization value did not change. The quantum yields (Φ) for the present
photodemagnetization were above one, e.g., $\Phi = 4.5$ (= 43 mJ cm^{-2} pulse $^{-1}$).
The irradiated sample returned to the LT phase by an annealing treatment
(3 K → 150 K → 3 K) with a relaxation at 120 K. The IR spectra before and
after one-shot-laser-pulse irradiation (532 nm, $P = 14$ mJ cm^{-2} pulse $^{-1}$) at
8 K was obtained. After irradiation, the Mn^{III} − NC − Fe^{II} peak at 2095 cm^{-1}
disappears and a sharp peak due to the Mn^{II} − NC − Fe^{III} peak at 2152 cm^{-1}
appears. Note that, in the case of $P < P'_{th}$, the IR spectra were not changed
by irradiation of several tens shots. An annealing treatment (8 K → 150 K →
8 K) returned the IR spectrum of the irradiated sample to the LT phase. From

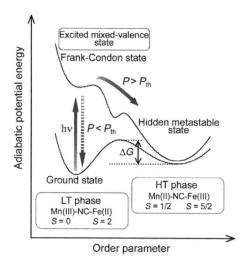

Fig. 1.24. Schematic illustration of the one-shot-laser-pulse-induced phase transition from the stable $Mn^{III} - NC - Fe^{II}$ phase to the hidden substable $Mn^{II} - NC - Fe^{III}$ phase

these results, we conclude that the present photo-demagnetization is caused by the photo-induced phase transition from the LT phase to the HT phase.

Temperature-induced phase transition between the LT and HT phases was observed in rubidium manganese hexacyanoferrate. In such a material with a bistability, a ground state can be converted to a hidden metastable state by the irradiation. Nasu et al. showed a simple scheme for a photo-induced phase transition using the adiabatic potential energy vs. order parameter (Fig. 1.24) [2]. In this scheme, the ground state is excited to the Franck–Condon state by irradiation. This Franck–Condon state proceeds to a hidden substable state through a structural change state or relaxes to the ground state. In our case, irradiating with pulsed-laser light excites the LT phase to a mixed-valance state between the $Mn^{III} - Fe^{II}$ and the $Mn^{II} - NC - Fe^{III}$ states. This mixed-valence state proceeds to the HT phase or relaxes to the initial LT phase. The produced HT phase can be maintained in the low temperature range since it is sufficiently separated from the LT phase by the thermal energy (ΔG). In addition, when the P value is larger than P_{th}', the excited state proceeds to the photo-produced HT phase as shown by the solid arrow in Fig. 1.24. In contrast, when $P < P_{th}'$, the excited state relaxes to the ground state as shown by the dotted arrow.

1.9.2 Reversible Photomagnetic Effect

In previous section, we have reported that irradiating with 532 nm light converts the LT phase to the photo-induced (PI) phase, which decreases its spontaneous magnetization. In this section, we have found that irradiating with a different wavelength of light recovers the PI phase in $Rb_{0.88}Mn[Fe(CN)_6]_{0.96} \cdot 0.5H_2O$ to the LT phase. Furthermore, neutron powder diffraction using an analog complex, $Rb_{0.58}Mn[Fe(CN)_6]_{0.86} \cdot 2.3H_2O$, has confirmed the magnetic ordering of the PI phase. Herein, we show the visible-light reversible changes in the electronic and magnetic properties of $Rb_{0.88}Mn[Fe(CN)_6]_{0.96} \cdot 0.5H_2O$, the neutron powder diffraction pattern of $Rb_{0.58}Mn[Fe(CN)_6]_{0.86} \cdot 2.3H_2O$, and the mechanism of the observed photo-reversible photomagnetism [77].

Photo-Reversible Changes in the IR Spectra and SQUID Measurement

Because a metal-to-metal charge transfer (MM'CT) band was observed at 420–540 nm in the ε spectrum of the LT phase (Fig. 1.5), the LT phase was irradiated with a CW diode green laser ($h\nu 1$; $\lambda = 532$ nm). Figure 1.25 shows the IR spectra before and after the light irradiations at 3 K. Before irradiating (Fig. 1.25a, black line), the LT phase possessed a broad peak due to $Mn^{III} - NC - Fe^{II}$ around 2100 cm^{-1}. Irradiating with $h\nu 1$ reduced the $Mn^{III} - NC - Fe^{II}$ peak, and created a sharp peak at 2153 cm^{-1}. The latter peak is assigned to the $Mn^{II} - NC - Fe^{III}$, which was also observed in the HT phase (2154 cm^{-1}). It is concluded that the PI phase after $h\nu 1$ irradiation has a valence state similar to the HT phase. Based on the knowledge that resonance due to the LMCT band was observed at 410 nm in the spectrum of the HT phase, this PI phase was irradiated with blue light ($h\nu 2$;

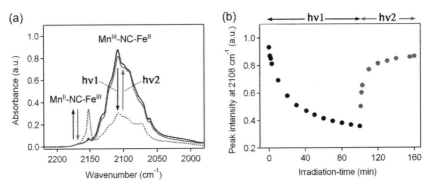

Fig. 1.25. Visible-light reversible change in the IR spectra of $Rb_{0.88}Mn[Fe(CN)_6]_{0.96} \cdot 0.5H_2O$. (a) Changes in the IR spectrum at 3 K by irradiating with $h\nu 1$ ($\lambda = 532$ nm: *black arrows*) before irradiation (*black line*), after $h\nu 1$ irradiation (*dotted line*), and $h\nu 2$ irradiation (*gray line*). (b) Peak intensity at 2108 vs. irradiation-time upon irradiating with $h\nu 1$ (*black circles*) and $h\nu 2$ (*gray circles*)

$\lambda = 410\pm30$ nm) from a filtered Xe lamp in order to investigate the photo-reversibility. Consequently, the $Mn^{II} - NC - Fe^{III}$ peak decreased and the $Mn^{II} - NC - Fe^{III}$ peak increased as shown in Fig. 1.25a. Figure 1.25b plots the peak intensities of $Mn^{II} - NC - Fe^{III}$ vs. irradiation-time. This photo-reversibility was repeatedly observed.

Next, we measured the photo-reversible change in magnetization *in situ* using SQUID equipment. The field cooled magnetization curve under an external magnetic field of 200 Oe showed that the LT phase is a ferromagnet with a Tc of 12 K (Fig. 1.26a, closed squares). Upon irradiating with hν1 at 3 K, the magnetization value decreased from 5600 to 700 G cm^3 mol^{-1} (Fig. 1.26a, open circles).

Successively irradiating the PI phase with hν2 increased the magnetization, which reached 4700 G cm^3 mol^{-1} (Fig. 1.26a, closed circles). The present photo-reversibility of the magnetization was repeatedly observed by alternately irradiating with hν1 and hν2 (Fig. 1.26b). The magnetization value after irradiating with hν2, which is shown as the closed circles in Fig. 1.26a, was smaller than that of initial value, suggesting that a photo-equilibrium state persists. To confirm the photo-equilibrium behavior, we investigated the photo-effect of the reverse process, that is, from the PI phase to the LT phase, using a different light (hν3 ; = 425$\pm$45 nm). Irradiating with hν3 increased the magnetization, which reached plateau- **Mag**$_{h\nu3}$ as shown in Fig. 1.27. Subsequent irradiation with hν2 further increased the magnetization, which reached plateau-**Mag**$_{h\nu2}$. This equilibrium behavior is due to a photo-stationary state between the photo-demagnetization (LT $\rightarrow$ PI phase) and the photo-induced magnetization (PI $\rightarrow$ LT phase).

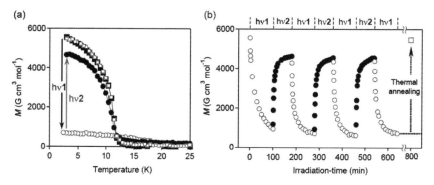

Fig. 1.26. Visible-light reversible photomagnetism in Rb$_{0.88}$Mn[Fe(CN)$_6$]$_{0.96}$· 0.5H$_2$O. (a) Magnetization vs. temperature curves at 200 Oe; before irradiating (■), after hν1 ($\lambda = 532$ nm, 30 mW cm^{-2}) irradiation for 100 min (○), after hν2 ($\lambda = 410$ nm, 13 mW cm^{-2}) irradiation for 80 min (●), and after the thermal annealing treatment of 180 K (□). (b) Magnetization vs. irradiation-time plot at 3 K by alternating with hν1 (○) and hν1 (●) light irradiation, and the magnetization value after a thermal treatment of 180 K (□)

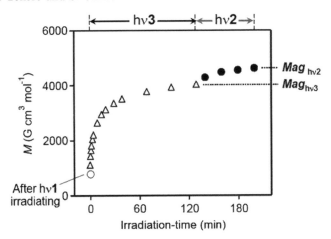

Fig. 1.27. Photo-stationary state between photodemagnetization and photo-induced magnetization. Magnetization vs. irradiation-time plot at 3 K upon irradiating with $h\nu3$ ($\lambda = 425 \pm 445$ nm, 22 mW cm^{-2}) (*open triangle*) and then $h\nu2$ ($\lambda = 410 \pm 30$ nm, 13 mW cm^{-2}) (*black circles*)

Magnetic Ordering of the Photo-Induced Phase

To determine the magnetic ordering of the PI phase, we performed neutron powder diffraction using an analog complex, $Rb_{0.58}Mn[Fe(CN)_6]_{0.86} \cdot 2.3H_2O$. A charge-transfer phase transition was not observed in the IR spectrum of $Rb_{0.58}Mn[Fe(CN)_6]_{0.86} \cdot 2.3H_2O$ when the sample was cooled to 3 K at a rate of –0.5 K (Fig. 1.28a). The $\chi_M^{-1} - T$ plot showed a negative Weiss temperature of –16 K, which was obtained by the least-square fitting in the temperature region of 150 - 320 K. The magnetization vs. temperature curve under an external field of 10 Oe exhibited an antiferromagnetic behavior with a Neel temperature (T_N) of 11.5 K. The magnetization vs. external magnetic field plots at 2 K showed a linear change (Fig. 1.28b). These magnetic data suggest that in $Rb_{0.58}Mn[Fe(CN)_6]_{0.86} \cdot 2.3H_2O$, the HT phase is maintained even at low temperature and the HT phase shows antiferromagnetism. Figure 1.28c shows the neutron powder pattern for $Rb_{0.58}Mn[Fe(CN)_6]_{0.86} \cdot 2.3H_2O$ at 30 K. Rietveld analysis showed that the crystal structure was tetragonal (P4/mmm) with lattice constants of $a = b = 7.424(6)$ Å and $c = 10.51(1)$ Å, which correspond to $a' = b' = 10.499$ Å and $c' = 10.51(1)$ Å in the frame of a cubic lattice. The interatomic distances of Fe and C in the ab plane (Fe $-$ C$_{ab}$) and along the c axis (Fe $-$ C$_c$) are 1.93(3) and 1.81(4) Å, respectively. The distances of Mn $-$ N$_{ab}$ and Mn $-$ N$_c$ are 2.18(2) and 2.18(4) Å, respectively.

Figure 1.29a and b shows the neutron powder diffraction patterns at 2 and 30 K, and the magnetic Bragg reflections as the difference in the patterns of 2 and 30 K, respectively. Analysis of the magnetic Bragg reflections suggests that this system is a layered antiferromagnet in which the magnetic coupling between the layers is antiferromagnetic. The spin arrangement as shown in

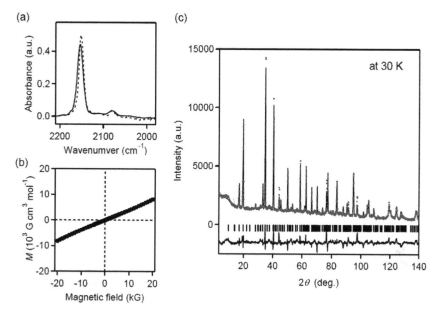

Fig. 1.28. Electronic state, magnetic property, and neutron powder diffraction pattern of $Rb_{0.58}Mn[Fe(CN)_6]_{0.86} \cdot 2.3H_2O$. (**a**) IR spectra at 300 K (*dotted line*) and 3 K (*solid line*). (**b**) Magnetization as a function of the external magnetic field at 2 K. (**c**) Neutron powder diffraction pattern at 30 K. *Gray dots, black pattern,* and *black line* are the observed plots, calculated pattern, and their difference, respectively. *Bars* represent the calculated positions of the Bragg reflections

Fig. 1.29d is a suitable configuration due to the following reason. The electronic state of Mn^{II} is a $3d^5$ high-spin state and hence, all the $3d$ orbitals are magnetic orbitals. In contrast, Fe^{III} is a $3d^5$ low-spin state, and thus, only one of the t_{2g} orbitals becomes a magnetic orbital. Rietveld analysis showed elongation of $Fe(CN)_6$ in the ab-plane, indicating that the d_{yz} and d_{zx} orbitals are more stabilized than the d_{xy} orbital due to backbonding of the cyanide ligand. Hence, the d_{xy} becomes the magnetic orbital of Fe^{III}. In this case, only the spin configuration shown in Fig. 1.29d is possible to be formed. The stick diagram of Fig. 1.29c, which was calculated by the layered antiferromagnet mentioned above, reproduced the observed data. Because the magnetic ordering of $Rb_{0.58}Mn[Fe(CN)_6]_{0.86} \cdot 2.3H_2O$ is considered to be the same as that of the PI phase in $Rb_{0.88}Mn[Fe(CN)_6]_{0.96} \cdot 0.5H_2O$, the PI phase should be a layered antiferromagnet .

Mechanism of Visible-Light Reversible Photomagnetism

The observed reversible photomagnetic e ffect can be explained by the scheme shown in Fig. 1.30. Irradiating with $h\nu 1$ excites the MM'CT ($Fe^{II} \rightarrow Mn^{III}$) band, which then excites the LT phase to photoexcited state **I**. Photoexcited

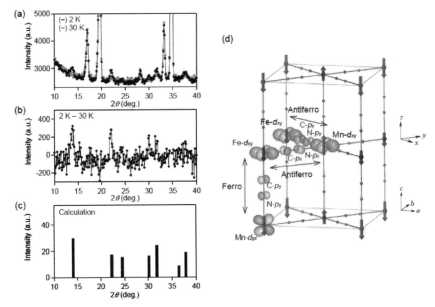

Fig. 1.29. (a) Neutron powder diffraction patterns at 2 K (*black line*) and 30 K (*gray line*). (b) Magnetic Bragg reflections as the difference in the neutron powder diffraction patterns at 2 and 30 K. (c) Calculated intensities of the magnetic Bragg reflections with an antiferromagnetic spin ordering. (d) Schematic illustration of the spin ordering. *Gray* and *Dark gray* arrows indicate the spins on Mn^{II} and Fe^{III}, respectively. From the view of the superexchange pathway, an antiferromagnetic coupling operates between Fe-d_{xy} and Mn-d_{xy} magnetic orbitals in the xy (ab) plane. In contrast, a ferromagnetic coupling operates between Fe-d_{xy} and all the d orbitals of Mn (here, Mn-d_{yz} is depicted) along the z (c) axis

state **I** proceeds to the PI phase, which has the same valence state as the HT phase. Thermal energy then suppresses the relaxation of the metastable PI phase to the stable LT phase. In contrast, the excitation of the LMCT ($CN^- \rightarrow Fe^{III}$) band of $[Fe(CN)_6]^{3-}$ by irradiating with $h\nu 2$ excites the PI phase to photoexcited state **III**, which then proceeds to the LT phase. The LT phase is a ferromagnet due to the ferromagnetic coupling between the $Mn^{III}(S = 2)$ sites, but the PI phase is an antiferromagnet. Hence, the magnetization value changes by optical switching between the LT phase and the PI phase.

A visible light-induced reversible photomagnetism between the ferromagnetic and antiferromagnetic phases is observed in a rubidium manganese hexacyanoferrate, $Rb_{0.88}Mn[Fe(CN)_6]_{0.96} \cdot 0.5H_2O$, by alternately irradiating with 532 and 410 nm lights. Optical switching from the LT phase to the PI phase occurs through a $Fe^{II} \rightarrow Mn^{III}$ MM'CT transition, causing photodemagnetization. In contrast, the reverse process is caused by an optical transition from the PI phase to the LT phase through a $CN^- \rightarrow Fe^{III}$ LMCT transition. The

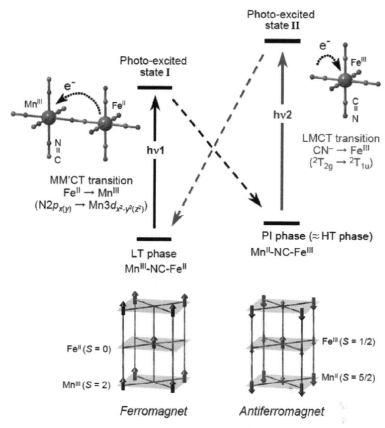

Fig. 1.30. Schematic illustration of the visible-light reversible photomagnetic effect in rubidium manganese hexacyanoferrate. Scheme for reversible charge-transfer between and (*upper*) and the spin ordering for the LT and PI phases (*lower*). LT phase is a ferromagnet due to ferromagnetic coupling between the sites, whereas the PI phase is an antiferromagnet. *Arrows* on the LT phase represent the spins of. *Large* and *small arrows* on the PI phase indicate the spins of and, respectively

existence of a photo-stationary state between the LT → PI phase and the PI → LT phase is also confirmed by the light source changing experiment. Although photomagnetism has been observed in some compounds, this is the first example of optical switching between a ferromagnet and an antiferromagnet.

1.10 Summary

In conclusion, a temperature-induced phase transition between the high-temperature (HT) and low-temperature (LT) phases is observed with a thermal hysteresis loop of 75 K in $RbMn[Fe(CN)_6]$. The charge transfer from Mn^{II}

to Fe^{III} that accompanies the Jahn–Teller effect on the $Mn^{III}N_6$ moiety explains this phase transition. By control the x of $Rb_x Mn[Fe(CN)_6]_{(x+2)/3} \cdot zH_2O$, we found that $Rb_{0.64}Mn[Fe(CN)_6]_{0.88} \cdot 1.7H_2O$ exhibits a surprisingly large thermal hysteresis loop of 138 K. A hidden stable phase of HT phase, which is experimentally observed in this system, is well explained by a SD model under thermal equilibrium condition. In addition, with non phase transition material of $Rb_{0.43}Mn[Fe(CN)_6]_{0.81} \cdot 3H_2O$, the *light-induced phase collapse* (LIPC) was realized. The LIPC is caused by blue-light irradiation inducing the transition from a *thermodynamically metastable phase* to a *hidden stable phase* in a material that does not undergo a thermal phase transition. Since the present phenomenon is driven only by the blue-light irradiation, it may provide a good strategy for the next generation of optical recording. As photo-induced phase transition at room temperature, the photoconversion from the LT to HT phases is observed inside the thermal hysteresis loop with a large Φ value of 38, by a one-shot-laser-pulse irradiation. This large yield and fast response will allow us to consider a new type of optical switching device. As photomagnetic effect at low temperature, the rapid- photodemagnetization has been observed by a one-shot-pulsed-laser light irradiation. With cw lights, a visible light-induced reversible ph otomagnetism between the ferromagnetic and antiferromagnetic phases is observed, by alternately irradiating with 532 nm and 410 nm lights. These temperature- and photo-induced phase transition phenomena are caused by: (1) the change in valence states on transition metal ions due to metal-to-metal charge-transfer and (2) the bistability due to the Jahn–Teller distortion of Mn^{III} ion.

Acknowledgements

The authors would like to thank Professor Yutaka Moritomo (University of Tsukuba) and Professor Kenji Ohoyama (Tohoku University) for the measurement of neutron powder diffraction . The present research is supported in part by a Grant-in-Aid for Young Scientists (S) from JSPS, a Grant for the GCOE Program "Chemistry Innovation through Cooperation of Science and Engineering", the photon Frontier Network Program from the MEXT, and PRESTO JST, Japan.

References

1. O. Kahn, *Molecular Magnetism* (VCH, New York, 1993)
2. K. Nasu, *Relaxations of Excited States and Photo-Induced Structual Phase Transitions* (Springer-Verlag, Berlin, 1997)
3. P. Gutlich, A. Hauser, H. Spiering, Angew. Chem. Int. Ed. Engl. **33**, 2024 (1994)
4. S. Ohkoshi, K. Hashimoto, J. Photochem. Photobio. C **2**, 71 (2001)
5. J.F. Letard, P. Guionneau, E. Codjovi, O. Lavastre, G. Bravic, D. Chasseau, O. Kahn, J. Am. Chem. Soc. **119**, 10861 (1997)

6. G.A. Renovitch, W.A. Baker, J. Am. Chem. Soc. **89**, 6377 (1967)
7. M. Sorai, Bull. Chem. Soc. Jpn. **74**, 2223 (2001)
8. K. Prassides, *Mixed Valency Systems, Applications in Chemistry, Physics and Biology* (NATO ASI, Kluwer, Dordrecht, 1991)
9. M.B. Robin, P. Day, Adv. Inorg. Chem. Radiochem. **10**, 247 (1967)
10. N.S. Hush, Prog. Inorg. Chem. **8**, 391 (1967)
11. S.B. Piepho, E.R. Krausz, P.N. Schatz, J. Am. Chem. Soc. **10**, 2996 (1978)
12. R.D. Cannon, L. Montri, D.B. Brown, K.M. Marshall, C.M. Elliot, J. Am. Chem. Soc. **106**, 2591 (1984)
13. H. Kitagawa, T. Mitani, Coord. Chem. Rev. **190**, 1169 (1999)
14. O.S. Jung, D.H. Jo, Y.A. Lee, B.J. Conklin, C.G. Pierpont, Inorg. Chem. **36**, 19 (1997)
15. N. Shimamoto, S. Ohkoshi, O. Sato, K. Hashimoto, Inorg. Chem. **41**, 678 (2002)
16. R.J. Zimmermann, Phys. Chem. Solids. **44**, 151 (1983)
17. T.J. Kambara, Phys. Soc. Jpn. **49**, 1806 (1980)
18. S. Ohnishi, S. Sugano, J. Phys. C **14**, 39 (1981)
19. A. Ludi, H.U. Gudel, Struct. Bonding (Berlin) **14**, 1 (1973)
20. M. Verdaguer, T. Mallah, V. Gadet, I. Castro, C. Helary, S. Thiebaut, P. Veillet, Conf. Coord. Chem. **14**, 19 (1993)
21. S. Ohkoshi, K. Hashimoto, Electrochem. Soc. Interface Fall **34**, (2002)
22. T. Mallah, S. Thiebaut, M. Verdaguer, P. Veillet, Science **262**, 1554 (1993)
23. W.R. Entley, G.S. Girolami, Science **268**, 397 (1995)
24. S. Ferlay, T. Mallah, R. Ouahes, P. Veillet, M. Verdaguer, Nature **378**, 701 (1995)
25. S. Ohkoshi, T. Iyoda. A. Fujishima, K. Hashimoto, Phys. Rev. B **56**, 11642 (1997)
26. S. Ohkoshi, A. Fujishima, K. Hashimoto, J. Am. Chem. Soc. **120**, 5349 (1998)
27. O. Hatlevik, W.E. Bushmann, J. Zhang, J.L. Manson, J.S. Miller, Adv. Mater. **11**, 914 (1999)
28. S.M. Holmes, G.S. Girolami, J. Am. Chem. Soc. **121**, 5593 (1999)
29. S. Ohkoshi, Y. Abe, A. Fujishima, K. Hashimoto, Phys. Rev. Lett. **82**, 1285 (1999)
30. S. Ohkoshi, K. Arai, Y. Sato, K. Hashimoto, Nat. Mater. 3, 857 (2004)
31. S. Margadonna, K. Prassides, A.N. Fitch, J. Am. Chem. Soc. **126**, 15390 (2004)
32. S.S. Kaye, J.R. Long, J. Am. Chem. Soc. **127**, 6506 (2005)
33. A.L. Goodwin, K.W. Chapman, C.J. Kepert, J. Am. Chem. Soc. **127**, 17980 (2005)
34. S. Ohkoshi, H. Tokoro, T. Matsuda, H. Takahashi, H. Irie, K. Hashimoto, Angew. Chem. Int. Ed. **3**, 857 (2007)
35. S. Ohkoshi, S. Yorozu, O. Sato, T. Iyoda, A. Fujishima, K. Hashimoto, Appl. Phys. Lett. **70**, 1040 (1997)
36. A. Bleuzen, C. Lomenech, V. Escax, F. Villain, F. Varret, C.C.D. Moulin, M. Verdaguer, J. Am. Chem. Soc. **122**, 6648 (2000)
37. O. Sato, S. Hayami, Y. Einaga, Z.Z. Gu, Bull. Chem. Soc. Jpn. **76**, 443 (2003); H. Tokoro, S. Ohkoshi, K. Hashimoto, Appl. Phys. Lett. **82**, 1245 (2003)
38. H. Tokoro, S. Ohkoshi, K. Hashimoto, Appl. Phys. Lett. **82**, 1245 (2003)
39. S. Ohkoshi, N. Machida, Z.J. Zhong, K. Hashimoto, Synth. Met. **122**, 523 (2001)
40. G. Rombaut, M. Verelst, S. Golhen, L. Ouahab, C. Mathoniere, O. Kahn, Inorg. Chem. **40**, 1151 (2001)

41. J.M. Herrera, V. Marvaud, M. Verdaguer, J. Marrot, M. Kalisz, C. Mathoniere, Angew. Chem. Int. Ed. **43**, 5468 (2004)
42. S. Ohkoshi, H. Tokoro, T. Hozumi, Y. Zhang, K. Hashimoto, C. Mathoniere, I. Bord, G. Rombaut, M Verelst, C.C.D. Moulin, F. Villain, J. Am. Chem. Soc. **128**, 270 (2006)
43. S. Ohkoshi, S. Ikeda, T. Hozumi, T. Kashiwagi, K. Hashimoto, J. Am. Chem. Soc. **128**, 5320 (2006)
44. S. Ohkoshi, Y. Hamada, T. Matsuda, Y. Tsunobuchi, H. Tokoro, Chem. Mater. **20**, 3048 (2008)
45. N. Yamada, E. Ohno, K. Nishiuchi, N. Akahira, M. Takao, J. Appl. Phys. **69**, 2849 (1991)
46. A.V. Kolobov, P. Fons, A.I. Frenkel, A.L. Ankudinov, J. Tominaga, T. Uruga, Nat. Mater. **3**, 703 (2004)
47. M. Wuttig, N. Yamada, Nat. Mater. **6**, 824 (2007)
48. S. Decurtins, P. Gutlich, C.P. Kohler, H. Spiering, A. Hauser, Chem. Phys. Lett. **105**, 1 (1984)
49. J.F. Letard, J.A. Real, N. Moliner, A.B. Gaspar, L. Capes, O. Cadpr, O. Kahn, J. Am. Chem. Soc. **121**, 10630 (1999)
50. S. Koshihara, Y. Tokura, T. Mikami, G. Saito, T. Koda, Phys. Rev. B **42**, 6853 (1990)
51. E. Collet, M.H. Lemee-Cailleau, M.B.L. Cointe, H. Cailleau, M. Wulff, T. Luty, S. Koshihara, M. Meyer, L. Toupet, P. Rabiller, S. Techert, Science **300**, 612 (2003)
52. M. Fiebig, K. Miyano, Y. Tomioka. Y, Tokura, Science **280**, 1925 (1998)
53. N. Takubo, I. Onishi, K. Takubo, T. Mizokawa, K. Miyano, Phys. Rev. Lett. **101**, 177403 (2008)
54. H. Tokoro, M. Shiro, K. Hashimoto, S. Ohkoshi, Z. Anorg. Allg. Chem. **633**, 1134 (2007)
55. S. Ohkoshi, H. Tokoro, M. Utsunomiya, M. Mizuno, M. Abe, K. Hashimoto, J. Phys. Chem. B **106**, 2423 (2002)
56. H. Tokoro, S. Ohkoshi, T. Matsuda, K. Hashimoto, Inorg. Chem. **43**, 5231 (2004)
57. H. Osawa, T. Iwazumi, H. Tokoro, S. Ohkoshi, K. Hashimoto, H. Shoji, E. Hirai, T. Nakamura, S. Nanao, Y. Isozumi, Solid State Commun. **125**, 237 (2003)
58. T. Yokoyama, H. Tokoro, S. Ohkoshi, K. Hashimoto, K. Okamoto, T. Ohta, Phys. Rev. B **66**, 184111 (2002)
59. S. Ohkoshi, T. Nuida, T. Matsuda, H. Tokoro, K. Hashimoto, J. Mater. Chem. **5**, 3291 (2005)
60. K. Kato, Y. Moritomo, M. Takata, M. Sakata, M. Umekawa, N. Hamada, S. Ohkoshi, H. Tokoro, K. Hashimoto, Phys. Rev. Lett. **91**, 255502 (2003)
61. H. Tokoro, S. Ohkoshi, T. Matsuda, T. Hozumi, K. Hashimoto, Chem. Phys. Lett. **388**, 379 (2004).
62. T. Nakamoto, Y. Miyazaki, M. Itoi, Y. Ono, N. Kojima, M. Sorai, Angew. Chem. Int. Ed. **40**, 4716 (2001)
63. H.M.J. Blote, Physica B **79B**, 427 (1975)
64. T. Matsumoto, Y. Miyazaki, A.S. Albrecht, C.P. Landee, M.M. Turnbull, M. Sorai, J. Phys. Chem. B **104**, 9993 (2000).
65. R.L. Carlin, *Magnetochemistry* (Springer, New York, 1986)
66. L.J.D. Jongh, A.R. Miedema, Adv. Phys. **23**, 1 (1974)

67. M.A. Subramanian, A.P. Ramirez, W.J. Marshall, Phys. Rev. Lett. **82**, 1558 (1999)
68. N. Ohmae, A. Kajiwara, Y. Miyazaki, M. Kamachi, M. Sorai, Thermochim. Acta **267**, 435 (1995)
69. B. Mayoh, P. Day, J. Chem. Soc. Dalton. **15**, 1483 (1976)
70. H. Tokoro, S. Miyashita, K. Kazuhito, S. Ohkoshi, Phys. Rev. B **73**, 172415 (2006)
71. C.P. Slichter, H.G. Drickamer, J. Chem. Phys. **56**, 2142 (1972)
72. K.P. Purcell, M.P. Edwards, Inorg. Chem. **23**, 2620 (1984)
73. H. Tokoro, S. Ohkoshi, Appl. Phys. Lett. **93**, 021906 (2008)
74. H. Tokoro, T. Matsuda, K. Hashimoto, S. Ohkoshi, J. Appl. Phys. **97**, 10M508 (2005)
75. H. Tokoro, K. Hashimoto, S. Ohkoshi, J. Magn. Magn. Mater. **310**, 1422 (2007)
76. S. Ohkoshi, H. Tokoro, K. Hashimoto, Coord. Chem. Rev. **249**, 1830 (2005)
77. H. Tokoro, T. Matsuda, T. Nuida, Y. Moritomo, K. Ohoyama, E.D.L. Dangui, K. Boukheddaden, S. Ohkoshi, Chem. Mater. **20**, 423 (2008)

2

Photoinduced Energy Transfer in Artificial Photosynthetic Systems

H. Imahori and T. Umeyama

2.1 Introduction

Artificial photosynthesis is a current topic of intensive investigations, both in order to understand the reactions that play a central role in natural photosynthesis as well as to develop highly efficient solar energy conversion systems and molecular optoelectronic devices [1–34]. Artificial photosynthesis is defined as a research field that attempts to mimic the natural process of photosynthesis. Therefore, the outline of natural photosynthesis is described briefly for the better understanding of artificial photosynthesis. Natural photosynthetic system is regarded as one of the most elaborate nanobiological machines [35, 36]. It converts solar energy into electrochemical potential or chemical energy, which is prerequisite for the living organisms on the earth. The core function of photosynthesis is a cascade of photoinduced energy and electron transfer between donors and acceptors in the antenna complexes and the reaction center. For instance, in purple photosynthetic bacteria (*Rhodopseudomonas acidophila* and *Rhodopseudomonas palustris*) there are two different types of antenna complexes: a core light-harvesting antenna (LH1) and peripheral light-harvesting antenna (LH2) [37–39]. LH1 surrounds the reaction center where charge separation takes place. The peripheral antenna LH2 forms two wheel-like structures: B800 with 9 bacteriochlorophyll *a* (Bchl *a*) molecules and B850 with 18 Bchl a molecules, which are noncovalently bound to two types of transmembrane helical α- and β-apoproteins. Carotenoids nearby the chlorophylls absorb sunlight in the spectral region where chlorophyll molecules absorb weakly and transfer the resultant excitation energy to the chlorophyll molecules via singlet-singlet energy transfer. The collected energy then moves from the LH2 to the LH1 in which the excitation energy migrates in the wheel-like arrays of chlorophylls of LH1 and LH2, and in turn is funneled into the chlorophyll dimer (special pair) in the reaction center [37–39]. Vectorial electron transfer event occurs along the array of chromophores embedded in the transmembrane protein, in a sequence of special pair, accessory chlorophyll, pheophytin, quinone A, and quinone B, yielding a long-lived, charge-separated state across the membrane with a quantum efficiency of ~100% [40, 41]. The final charge-separated state eventually results in the production of adenosine triphosphate (ATP) via the proton gradient generated across the membrane by the proton pump from the cytoplasm to the periplasm using the reduced quinone [35–41].

Light-harvesting systems also disclose a great variety in their structures. For instance, in purple bacteria, more than 9 chlorophylls are arranged in symmetric ring-like structures (*vide supra*) [37–39], while in green bacteria a large number of chlorophylls are organized into rod-like aggregates without the help of proteins [42]. On the other hand, chlorophyll aggregate in photosystem (PS) I of cyanobacteria and higher plants exhibits a rather random array forming two-dimensional (2D) structures, which surround the reaction center [43–45]. Such complex structure and diversity in natural light-harvesting systems have made it difficult to uncover the close structure-function relationship in the light-harvesting systems. In this context, various chromophore arrays, especially porphyrins, which are assembled covalently [46–97] or noncovalently [98–132], have been synthesized to shed light on the light-harvesting processes.

In this review we highlight our recent achievements relating to photoinduced energy transfer in artificial photosynthesis. In particular, the emphasis lies on self-assembled multiporphyrin arrays that are highly promising materials for photocatalysts, organic solar cells, and molecular optoelectronic devices throughout our studies.

2.2 Two-Dimensional Multiporphyrin Arrays

2.2.1 Self-assembled Monolayers of Porphyrins on Gold Electrodes

Covalently linked multiporphyrin arrays bearing more than ~10 porphyrin units are superior to the corresponding porphyrin monomers and oligomers with respect to the structural control and light-harvesting properties, but the synthetic difficulty makes it difficult to employ such covalently linked multiporphyrin arrays in terms of future practical applications [46–97]. Another promising approach for achieving these goals is the self-assembly of porphyrin bearing molecular recognition units. These porphyrin self-assemblies are prepared easily, but often afford less complete structural control and stability [98–132]. We have focused on self-assembled monolayers (SAMs) of porphyrins on flat gold substrates, because they can provide densely-packed, highly ordered structures of porphyrins on 2D gold electrodes suitable for developing artificial photosynthetic systems [98–108].

Systematic studies on the structure and photoelectrochemical properties of the SAMs of porphyrin disulfide dimers **1** on gold electrodes (denoted as Au/**1**) were performed to examine the effects of the spacer length, as shown in Fig. 2.1 [98, 100, 104]. In the molecular design of porphyrin disulfide dimers **1**, six *t*-butyl groups were introduced into the *meso*-phenyl rings of the

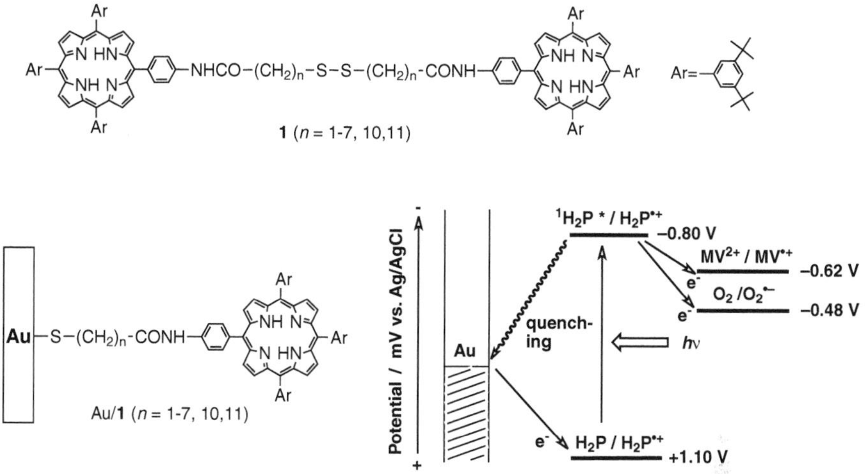

Fig. 2.1. Porphyrin disulfide dimers **1**, self-assembled monolayers of porphyrins on a gold electrode Au/**1**, and photocurrent generation diagram for Au/**1**/MV^{2+}/Pt device

porphyrin moieties to increase the solubility in organic solvents and to suppress the quenching of the porphyrin excited states in the monolayers owing to the porphyrin aggregation (Fig. 2.1) [104]. The structure of the SAMs was investigated using ultraviolet-visible absorption spectroscopy in transmission mode, cyclic voltammetry, ultraviolet-visible ellipsometry, and fluorescence spectroscopy [104]. These measurements revealed that the SAMs tend to form highly ordered structures on the gold electrode with increasing the methylene spacer length, reaching to surface coverage up to 1.5×10^{-10} mol cm^{-2} (110 Å^2 molecule^{-1}) [104]. The adjacent porphyrin rings are likely to adopt J-aggregate-like partially stacked structures in the monolayer [104]. The porphyrin ring plane in the monolayer with an even number ($n = 2,4,6,10$) of the methylene spacer (-(CH$_2$)$_n$-) is tilted significantly to the gold surface, while the porphyrin with an odd number ($n = 1,3,5,7,11$) of the methylene spacer takes a nearly perpendicular orientation to the gold surface [104].

The photoelectrochemical experiments on the gold electrodes modified with **1** were carried out in an argon-saturated Na$_2$SO$_4$ aqueous solution containing methyl viologen (MV^{2+}) as an electron carrier using a platinum wire counter electrode and an Ag/AgCl reference electrode (denoted as Au/**1**/MV^{2+}/Pt) [104]. An increase in the cathodic electron flow was observed with increasing the negative bias (0.7 to − 0.2 V) to the gold electrode [104]. This implies that vectorial electron transfer takes place from the gold electrode to the counter electrode through the SAM and the electrolyte. With increasing the spacer length, the adsorbed photon-to-current efficiency (APCE) of the photocurrent generation in the Au/**1**/MV^{2+}/Pt device was increased in a zigzag fashion to reach a maximum of 0.34% ($n = 6$) and then decreased slightly. Such dependence of the APCE value on the spacer length can be rationalized by the competition between electron and energy transfer quenching of the porphyrin singlet excited state as illustrated in Fig. 2.1. Photoirradiation of the modified electrode results in electron transfer from the porphyrin singlet excited state ($E^{0*}_{ox} = -0.80$ V vs. Ag/AgCl) to MV^{2+} ($E^{0}_{red} = -0.62$ V vs. Ag/AgCl) or O$_2$ ($E^{0}_{red} = -0.48$ V vs. Ag/AgCl) [104]. The reduced electron carriers diffuse to release electrons to the platinum electrode, whereas the resultant porphyrin radical cation (H$_2$P$^{\bullet+}$: +1.10 V vs Ag/AgCl) captures electrons from the gold electrode, generating the cathodic electron flow [104]. However, the energy transfer quenching of the porphyrin excited singlet state by the gold surface is competitive, judging from the extremely short fluorescence lifetimes (τ) of the porphyrins (10– 40 ps) on the gold surface compared with the values (1–10 ns) on quartz or semiconductor surfaces (vide infra) [104, 109, 111]. The electronic coupling between the porphyrin and the electrode is reduced with increasing the spacer length, leading to less energy transfer quenching of the porphyrin excited singlet state by the gold surface. On the other hand, an increase in the separation distance between the gold electrode and the H$_2$P$^{\bullet+}$ slows down the electron transfer from the gold electrode to the H$_2$P$^{\bullet+}$. Thus, the opposite effect as a function of the spacer length may be responsible for a nonlinear dependence of the APCE value on the spacer length (*vide supra*). These results manifest that the optimization of each process is vital to achieve efficient photocurrent generation in the SAMs of photoactive chromophores on gold electrodes.

To realize efficient photoinduced energy transfer on a gold electrode, it is essential to integrate suitable energy donor and acceptor on the gold surface. As the first attempt, mixed SAMs of pyrene and porphyrin on gold electrodes were prepared to address the possibility of photoinduced energy transfer on the surface. Monolayers of a mixture of porphyrin disulfide dimer **1** ($n = 11$) and pyrene disulfide dimer **2** were formed by the coadsorption of **1** ($n = 11$) and **2** onto Au(111) mica substrates (denoted as Au/**1**–**2**), as depicted in Fig. 2.2 [102, 106].

The electrochemical measurements of Au/**1** and Au/**2** in CH$_2$Cl$_2$ containing 0.1 M n-Bu$_4$NPF$_6$ electrolyte suggested the formation of well-packed structures of **1** and **2** on the gold surfaces [102, 106]. In the mixed SAMs, however, the waves arising from the first oxidations of the porphyrin and pyrene moieties were too broad to determine the adsorbed amounts of **1** and **2** in Au/**1**–**2** accurately. The ratio of **1** : **2** in Au/**1**–**2**, estimated from the absorption spectrum on the gold surface, is significantly lower than the value in CH$_2$Cl$_2$. The strong $\pi-\pi$ interaction of the pyrene moieties in comparison with the relatively weak interaction between the porphyrin moieties, because of the bulky t-butyl groups, may be attributed to the preference of the adsorption of **2** over **1** on the gold surface. In addition, the fact that the pyrene molecule occupies about half the surface area of the porphyrin would lead to a thermodynamic preference for pyrene adsorption, since displacement of a porphyrin for two pyrenes results in an extra S–Au interaction [102, 106].

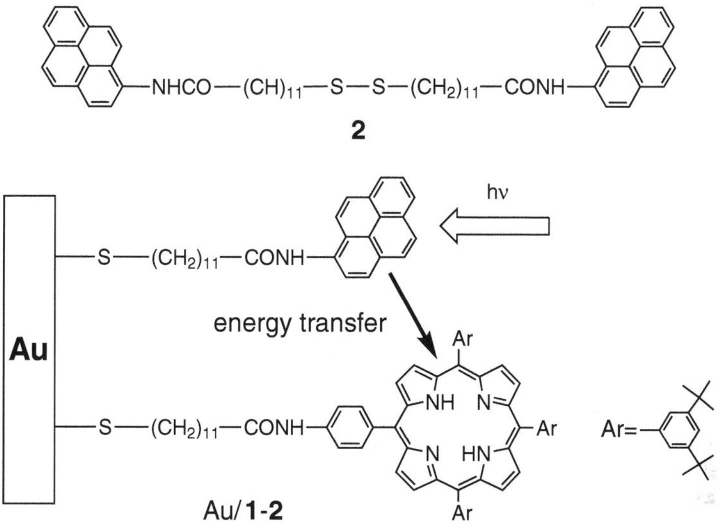

Fig. 2.2. Pyrene disulfide dimer **2** and mixed self-assembled monolayers of porphyrin and pyrene on a gold electrode Au/**1**–**2**

To probe a singlet-singlet energy transfer from the pyrene excited singlet state to the porphyrin moiety in the mixed SAMs, time-resolved, single-photon counting fluorescence measurements were made for Au/**1**–**2** as well as **1** and **2** in CH_2Cl_2 with an excitation wavelength of 280 nm, where the light is mainly absorbed by the pyrene moiety [102, 106]. The decay of the fluorescence intensities at $\lambda_{obs} = 385$ and 720 nm, arising from the pyrene and the porphyrin singlet excited states, respectively, could be monitored. The decay curve could be fitted as a single exponential except for the case of **2** at 385 nm in CH_2Cl_2. The fluorescence lifetimes of Au/**2** at 385 nm (23 ps) and Au/**1** at 720 nm (40 ps) are much shorter than those of **2**(7.4 ns (30%), 3.2 ns (70%)) and **1** (8.1 ns) in CH_2Cl_2. This indicates that the excited singlet states of the pyrene and the porphyrin moieties in the SAMs are efficiently quenched by the gold surface via energy transfer. However, it should be noted here that the fluorescence lifetime of the pyrene moiety in Au/**1**–**2** at 385 nm decreases with increasing the ratio of the porphyrin to the pyrene. The fluorescence lifetime of the porphyrin moiety in Au/**1**–**2** at 720 nm is also decreased with increasing the ratio of the porphyrin to the pyrene. Thus, we can conclude that efficient energy transfer occurs from the pyrene excited singlet state to the porphyrin, followed by energy migration among the porphyrins, which can compete with the energy transfer quenching by the gold surface [102, 106].

Although efficient singlet-singlet energy transfer occurs from the pyrene excited singlet state to the porphyrin in the mixed SAMs of pyrene and porphyrin on the gold surface (*vide supra*), pyrene can absorb light in the ultraviolet region solely ($\lambda_{max} = 337$ nm), thereby making it impossible to

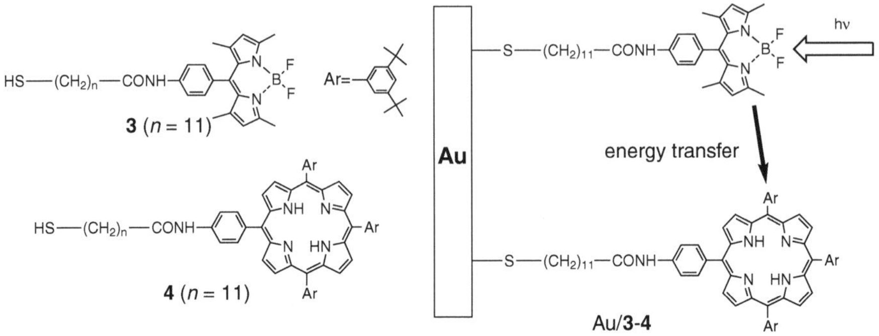

Fig. 2.3. Boron dipyrrin **3** ($n = 11$) and porphyrin alkanethiol **4** ($n = 11$) and mixed self-assembled monolayers of boron dippyrin and porphyrin on a gold electrode Au/**3**–**4**

collect light in the visible region (>400 nm). Boron dipyrrin thiol **3** was then chosen as an improved light-harvesting molecule to achieve efficient energy transfer from the boron dipyrrin excited singlet state ($^1B^*$) in **3** to the porphyrin (H_2P) in **4** in the mixed SAMs that were prepared from **3** and porphyrin alkanethiol **4** (Fig. 2.3) [106]. The boron-dipyrrin dye exhibits a moderately strong absorption band in the visible region around 500 nm ($\sim 10^5$ $M^{-1}cm^{-1}$) and a relatively long singlet excited-state lifetime (~ 5 ns) [106]. Taking into account the fact that the porphyrin moiety in **4** absorbs strongly in the blue ($\sim$420 nm) and weakly in the green region, an incorporation of the boron-dipyrrin pigments **3** into a SAM of **4** (denoted as Au/**4**) allows us to enhance the absorption properties in the green-region as well as the blue region. More importantly, the fluorescent emission ($\sim$510 nm) from the boron dipyrrin overlaps well with the absorption of Q bands (500–650 nm) of the porphyrin. Thus, an efficient singlet-singlet energy transfer from the ($^1B^*$) in **3** in **3** to the H_2P in **4** is anticipated to occur in the mixed SAMs of **3** and **4** on the gold surface (denoted as Au/**3**–**4**).

Actually, based on the energy diagram, energy transfer is expected to take place from $^1B^*$ in **3** to H_2P in **4**, followed by intermolecular electron transfer from the resulting $^1H_2P^*$ to diffusing electron carriers such as O_2 and MV^{2+} in the electrolyte, which eventually gives electrons to the counter electrode. On the other hand, the gold electrode gives electrons to the $H_2P^{\bullet+}$, generating vectorial electron flow from the gold electrode to the counter electrode through the SAM and the electrolyte [106]. Thus, the present system can mimic both photosynthetic energy transfer and electron transfer in the mixed SAM.

Cyclic voltammogram measurements indicated the formation of densely-packed monolayers of **3** and **4** on the gold surfaces, as in the cases of **1** and **2** [106]. The estimated ratio of **3** : **4** in the mixed SAMs is significantly higher than the value in the solution. The preference of the adsorption of **3** over **4** on the gold surface may also result from the strong $\pi - \pi$ interaction of the planar boron-dipyrrin moieties against relatively weak interaction between the porphyrin moieties as a result of the bulky t-butyl groups and the thermodynamic preference for boron-dipyrrin adsorption against the porphyrin that possesses the larger occupied area (*vide supra*) [106]. Cyclic voltammogram measurements indicated the formation of densely-packed monolayers of **3** and **4** on the gold surfaces, as in the cases of **1** and **2** [106]. The estimated ratio of **3** : **4** in the mixed SAMs is significantly higher than the value in the solution. The preference of the adsorption of **3** over **4** on the gold surface may also result from the strong $\pi - \pi$ interaction of the planar boron-dipyrrin moieties against relatively weak interaction between the porphyrin moieties as a result of the bulky t-butyl groups and the thermodynamic preference for boron-dipyrrin adsorption against the porphyrin that possesses the larger occupied area (*vide supra*) [106].

Efficient energy transfer from $^1B^*$ in **3** to H_2P in **4** was confirmed by the fluorescence spectrum of Au/**3-4** that reveals fluorescent emission from the porphyrin moiety solely ($\lambda_{max} = 650, 720$ nm) irrespective of an excitation wavelength ($\lambda_{ex} = 510$ or 420 nm) [106]. The energy transfer efficiency from $^1B^*$ in **3** to H_2P in **4** rises with rising the ratio of the porphyrin to the boron dipyrrin, to reach a maximum value of 100% at a ratio of **3** : **4** = 69 : 31. Under the optimized conditions, the excitation spectrum of Au/**3-4** with the fixed emission wavelength ($\lambda_{em} = 650$ nm) matches the absorption spectrum of Au/**3-4**. This unambiguously corroborates that efficient energy transfer takes place from $^1B^*$ in **3** to H_2P in **4** in the mixed SAMs on the gold surfaces.

The photoelectrochemical measurements were carried out using the SAMs of **3** and/or **4** on the gold electrodes in the standard three-electrode arrangement under the optimized conditions [electrolyte solution: O_2-saturated 0.1 M Na_2SO_4 solution containing 30 mM MV^{2+}] (denoted as Au/**3-4**/MV^{2+}/Pt) [106]. Unfortunately, the APCE values of the photocurrent generation in the Au/**3**/MV^{2+}/Pt, Au/**4**/MV^{2+}/Pt, and Au/**3-4**/MV^{2+}/Pt devices are similar under the same conditions as a consequence of comparable performance of the Au/**3**/MV^{2+}/Pt and Au/**4**/MV^{2+}/Pt devices. Accordingly, we could not obtain unambiguous evidence for the photocurrent generation resulting from photoinduced energy transfer from $^1B^*$ in **3** to the H_2P in **4** in the mixed SAMs of **3** and **4** on the gold electrode.

To enhance the energy transfer-assisted photocurrent generation, it is crucial to incorporate efficient photocurrent generation molecules into a SAM of light-harvesting molecules. Thus, ferrocene (Fc)-porphyrin (H_2P)-C_{60} triad **5** [103, 105] was employed for the boron-dipyrrin SAM to improve the quantum yield of photocurrent generation (Fig. 2.4). The triad thiol **5** was designed to reveal photoinduced electron transfer from the $^1H_2P^*$ to the C_{60}, followed by the efficient electron transfer from the ferrocene to the resulting $H_2P^{\bullet+}$, to yield the final charge-separated state, Fc^+-H_2P-$C_{60}^{\bullet-}$, as demonstrated in solutions [133, 134]. The $C_{60}^{\bullet-}$ moiety in the charge-separated state gives electrons to electron carriers such as MV^{2+} and O_2 in the electrolyte, whereas electrons are shifted from the gold electrode to the Fc^+ moiety, resulting in

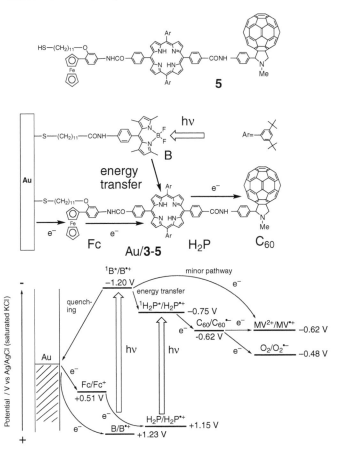

Fig. 2.4. Ferrocene-porphyrin-fullerene triad **5**, a self-assembled monolayer of boron dipyrrin and ferrocene-porphyrin-fullerene triad on a gold electrode Au/**3–5**, and photocurrent generation diagram for the Au/**3–5**/MV^{2+}/Pt device

the efficient cathodic electron flow, as displayed in Fig. 2.4. In addition, the emission from the boron dipyrrin in **3**, which exhibits better light-harvesting properties around 500 nm than **5**, matches well with the absorption of the porphyrin in **5**, as seen in the case of mixed SAMs of **3** and **4**. Thus, an efficient energy transfer from the ^{1}B* in **3** to the H$_2$P in **5** would take place in the mixed SAMs of **3** and **5** on the gold surface (Fig. 2.4). Overall, it is anticipated that the mixed SAMs of **3** and **5** can lead to efficient photocurrent generation, which also mimics the light-harvesting and charge separation in photosynthesis.

The amounts of **3** and **5** on the gold surface were systematically altered by the competitive coadsorption onto the gold surface from CH$_2$Cl$_2$ solutions containing various molar ratios of **3** and **5** (molar ratio of **3** : **5** = 100 : 0; 75 : 25;

50 : 50; 25 : 75; 10 : 90; 0 : 100) [106]. From the cyclic voltammetric and absorption spectral measurements, the ratio of **3** : **5** in Au/**3**–**5** was found to be comparable to that in CH_2Cl_2. In contrast to the cases of Au/**1**–**2** and Au/**3**–**4** (*vide supra*), no significant preference of the adsorption of **3** over **5** on the gold surface was observed, indicating that $\pi - \pi$ interaction of the boron-dipyrrin moieties is similar to that of **5** that contains both the porphyrin and fullerene moieties.

The photoelectrochemical measurements were conducted using the mixed SAMs of **3** and **5** on the gold electrodes in the three electrode arrangement (denoted as Au/**3**–**5**/MV^{2+}/Pt) [106]. The APCE values of the Au/**3**–**5**/MV^{2+}/Pt device, determined based on the absorption of the porphyrin and the antenna molecules at 430 and 510 nm, increase with increasing the content of **5** in the SAMs. The energy transfer efficiency from $^1B^*$ in **3** to H_2P in **5** may also raise with raising the content of **5** in the SAMs to reach maximum APCE values of 21±3% at 430 nm and 50±8% at 510 nm with a ratio of **3** : **5** = 37 : 63. The incident photon-to-current efficiencies (IPCE) of the Au/**3**–**5**/MV^{2+}/Pt device at 510 and 430 nm were also determined as 0.6 and 1.6%, respectively [106]. Formation of the charge-separated state (i.e., Fc^+-H_2P-$C_{60}{}^{\bullet-}$) in **5** following the energy and electron transfer steps in Fig. 2.4 has been well established by the time-resolved transient absorption studies of the triad molecule together with the fluorescence lifetime measurements, although the small absorbance of the present system has precluded the direct detection of the charge-separated state within the monolayer [106,133,134]. The APCE value (50±8%) at 510 nm is much higher than those of the Au/**3**/MV^{2+}/Pt, Au/**3**–**4**/MV^{2+}/Pt, and Au/**5**/MV^{2+}/Pt devices at 510 nm, and this is the highest value ever reported for photocurrent generation at monolayer-modified metal electrodes using donor-acceptor linked molecules [117, 118, 135, 136]. The coexistence of **3** as an antenna molecule in the Au/**3**–**5**/MV^{2+}/Pt device has enabled the utilization of the longer wavelength (510 nm) more efficiently than the device without **3**. It should be noted here that the APCE value of the Au/**3**–**5**/MV^{2+}/Pt device at 510 nm is two times as large as that at 430 nm. The fluorescence lifetime measurements of SAMs of porphyrins on the gold surface indicate that $^1H_2P^*$ is strongly quenched by the gold surface through energy transfer [104]. Thus, the higher APCE value at 510 nm may stem from the difference in quenching efficiency of the S_1 and S_2 states of the porphyrin in **5** by the gold electrode and/or surface plasmon effect due to the gold surface.

2.2.2 Self-assembled Monolayers of Porphyrins on ITO Electrodes

As described in Sect. 2.2.1, we have successfully achieved photosynthetic electron and energy transfer on the gold electrode modified with SAMs of porphyrins and related photoactive chromophores. However, strong energy transfer quenching of the porphyrin excited singlet state by the flat gold electrode has precluded achievement of a high quantum yield for charge separation on the surface as attained in natural photosynthesis. To surmount such an energy transfer quenching problem, indium-tin oxide (ITO) with high optical transparency (>90%) and electrical conductivity ($\sim 10^4$ Ω cm) seems to be highly attractive as an electrode. The quenching of the porphyrin excited singlet state on the surface may be suppressed because the conduction band (CB) of ITO is higher than the energy level of the porphyrin excited singlet state. Despite these advantages, development of SAMs on the ITO electrode has been rather limited in that their chemical modification requires carefully controlled conditions that have been difficult to achieve [111]. As such, the substituent effects of porphyrins in SAMs on ITO have not been fully addressed.

We examined the effects of bulkiness in porphyrin SAMs on the structure and photoelectrochemical properties [114]. 5,10,15,20-Tetraphenylporphyrin (TPP) and 5,10,15,20-tetraphenylpoprhyrin with bulky *tert*-butyl groups at the *meta* positions of the *meso*-phenyl groups (TBPP) were covalently linked to the ITO surface (denoted as ITO/6 and ITO/7), respectively, as illustrated in Fig. 2.5. The ultraviolet-visible absorption, steady-state fluorescence, and cyclic voltammetry measurements for the porphyrin SAMs revealed that the interaction between the porphyrins without bulky *tert*-butyl groups is much larger than that of the porphyrins with bulky *tert*-butyl groups. Photoelectrochemical measurements were carried out in a nitrogen-saturated Na_2SO_4 aqueous solution containing triethanolamine (TEA) as an electron sacrificer

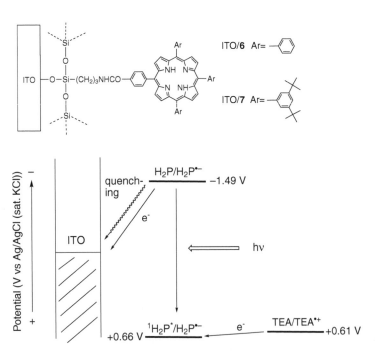

Fig. 2.5. Self-assembled monolayers of porphyrins on ITO electrodes ITO/6 and ITO/7 and photocurrent generation diagram for the ITO/6 or 7/TEA/Pt device

in the three electrode arrangement using the ITO/**6** and ITO/**7** electrodes (denoted as ITO/**6**/TEA/Pt and ITO/**7**/TEA/Pt) [114]. Surprisingly, the APCE value of the ITO/**6**/TEA/Pt device (2.2±0.9%) is virtually the same as that of ITO/**7**/TEA/Pt device (3.4±0.6%), although there is a large difference in the interaction between porphyrins in ITO/**6** and ITO/**7** electrodes owing to the steric hindrance of the bulky t-butyl groups. The fluorescence lifetimes of ITO/**6** [τ= 1.3 ns (40%), 5.9 ns (60%)] and ITO/**7** [τ= 1.9 ns (29%), 8.2 ns (71%)] are moderately reduced relative to TPP (τ= 10 ns) and TBPP (τ= 10 ns) in THF. This means that the quenching of porphyrin excited singlet state on ITO is remarkably suppressed compared with the intensive quenching on flat gold electrodes (τ= 1−40 ps) [104, 109, 111]. It is noteworthy that the τ values of ITO/**6** are largely similar to those of ITO/**7**. The results are in marked contrast with severe self-quenching of the porphyrin excited singlet state in conventional molecular assemblies such as Langmuir-Blodgett films [114]. The picosecond fluorescence anisotropy decay measurements for ITO/**6** and ITO/**7** suggested the occurrence of fast energy migration between porphyrin moieties in the SAMs [114]. The two-dimensional, densely-packed structure of the porphyrins in SAMs is responsible for the long-lived excited singlet state, which resembles the antenna function of photosystem I in cyanobacteria [43–45]. This conclusion is important for further development of porphyrin SAMs exhibiting antenna function, since we can densely pack porphyrin molecules on a 2D electrode surface where fast energy migration takes place between the porphyrins without losing the excitation energy.

2.3 Three-Dimensional Porphyrin Arrays

2.3.1 Self-assembled Monolayers of Porphyrins on Metal Nanoparticles

To overcome the problem, novel artificial light-harvesting systems, which remarkably enhance the light-harvesting properties, should be exploited to combine with charge separation system on an ITO electrode that also suppresses the undesirable energy transfer quenching. Metal nanoparticles, which can provide three-dimensional (3D) nanospace on the surface, are highly promising as nano scaffolds for antenna molecules [137–145]. In particular, alkanethiolate-monolayer protected gold nanoparticles are stable in air, soluble in common organic solvents, therefore being capable of facile modification with other functional thiols through exchange reactions or by couplings and nucleophilic substitutions [139, 140]. Therefore, construction of the 3D architectures of porphyrin-modified gold nanoparticles, which have large surface area, on ITO electrode would improve the light-harvesting efficiency compared with the 2D porphyrin SAMs. Furthermore, the interaction of porphyrin excited singlet state with gold nanoparticles would be reduced significantly,

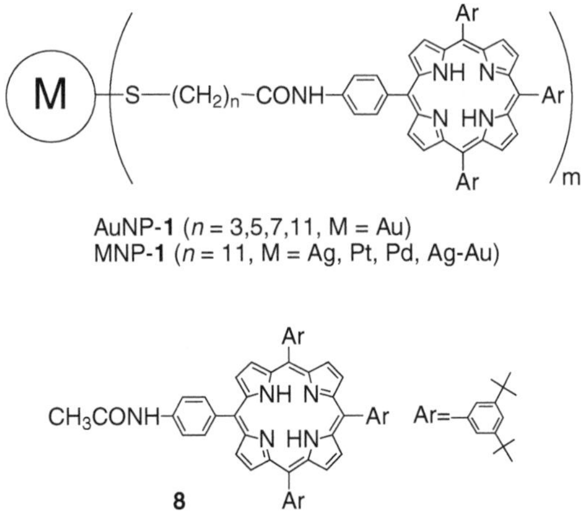

AuNP-1 (n = 3,5,7,11, M = Au)
MNP-1 (n = 11, M = Ag, Pt, Pd, Ag-Au)

Fig. 2.6. Porphyrin-modified metal nanoparticles MNP-1 ($n = 3,5,7,11$, M = Au, Ag, Pt, Pd, Ag–Au) and porphyrin reference **8**

relative to bulk gold surface, due to the "quantum effect" [141–145]. In this context, multiporphyrin monolayer-modified gold nanoparticles AuNP-1 ($n = 3,5,7,11$) were prepared as a new type of artificial photosynthetic materials (Fig. 2.6) [146–154]. The photophysical and electrochemical properties of AuNP-1 ($n = 3,5,7,11$) are compared to the corresponding 2D porphyrin SAM (Au/1 ($n = 3,5,7,11$)), as shown in Fig. 2.1 [146, 148]. For instance, AuNP-1 ($n = 11$) was directly synthesized by reduction of $AuCl_4^-$ with $NaBH_4$ in toluene containing the corresponding porphyrin disulfide dimer **1** or porphyrin alkanethiol **4** to avoid incomplete functionalization. AuNP-1 ($n = 11$) was purified repeatedly by gel permeation chromatography and characterized by ^{1}H NMR, UV-visible and fluorescence spectroscopies, electrochemistry, elemental analysis, and transmission electron microscopy (TEM) [146, 148].

The mean diameter of the gold core determined by TEM was 2.1 nm (with a standard deviation $\sigma = 0.3$ nm) for AuNP-1 ($n = 11$), which is comparable to the value obtained for alkanethiolate-protected gold nanoparticle under the same experimental conditions [146, 148]. Taking the gold core as a sphere, the model predicts that the core of AuNP-1 ($n = 11$) contains 280 gold atoms, of which 143 lie on the gold surface. Given the values for elemental analysis of AuNP-1 ($n = 11$), there are 57 porphyrin alkanethiolate chains on the gold surface for AuNP-1 ($n = 11$). It should be emphasized here that the molecular weight of AuNP-1 ($n = 11$) is estimated as 120,000, which is one of the highest values for multiporphyrin arrays with well-defined structure [21, 78–81]. The coverage ratio of porphyrin alkanethiolate chains of AuNP-1 ($n = 11$) to surface gold atoms (γ) is determined as 40%, which is remarkably increased relative to the coverage ratio ($\gamma = 6.5\%$) of 2D porphyrin SAM Au/1 ($n = 11$). In other words, the light-harvesting properties of the 3D system are much improved compared with those of the 2D system. ^{1}H-NMR, cyclic voltammetry, and absorption measurements of AuNP-1 ($n = 11$) revealed that the porphyrin environment of AuNP-1 ($n = 11$) is virtually the same as that of porphyrin reference 8 in solution and is less perturbed than that of Au/1 ($n = 11$) [146, 148].

To establish the excited state deactivation pathways, nanosecond transient absorption spectra were recorded for AuNP-1 ($n = 11$) and 8 in benzonitrile [148]. AuNP-1 ($n = 11$) and 8 exhibit characteristic absorption arising from the porphyrin excited triplet state, but the intensity of transient absorption for AuNP-1 ($n = 11$) is much lower than that of 8 under the same experimental conditions. This implies that most of the porphyrin excited singlet state on the gold nanoparticles is quenched by the metal surface, whereas residual porphyrin excited triplet state is generated *via* the intersystem crossing from the unquenched porphyrin excited singlet state. Picosecond transient absorption spectra were also taken for AuNP-1 ($n = 11$) in benzonitrile (Fig. 2.7) [148]. Immediately after the excitation of AuNP-1 ($n = 11$), the transient absorption arising from the porphyrin excited singlet state appears. The decay rate constant of the band ($k_q = 7.7 \times 10^9 \text{s}^{-1}$ at 460 nm) agrees well with the values for the rise of a hot band due to the surface plasmon around 600 nm [155] as well as for a short component of the fluorescence decay (Inset of Fig. 2.7). There is no evidence for the formation of the porphyrin radical cation. These results suggest that the porphyrin excited singlet state in the present systems is quenched by the metal surface via energy transfer rather

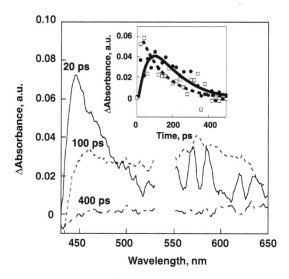

Fig. 2.7. Picosecond transient absorption spectra of AuNP-1 ($n = 11$) in benzonitrile as a function of the time delay between the pump and probe laser beams (20, 100, 400 ps) at an excitation of 540 nm. The *inset* displays the time profiles at 460 nm (*dotted curve with white squares*) and 600 nm (*solid curve with black circles*)

than electron transfer [148]. The fluorescence of the 3D porphyrin AuNP-1 ($n = 11$) exhibited a double exponential decay [0.13 ns (93%), 9.6 ns (7%)]. The lifetime of the longer-lived, minor component of AuNP-1 ($n = 11$) in benzonitrile is close to that of **8** ($\sim$10 ns). This minor component may result from different ligation sites (vertex, edge, terrace, and defect) on the truncated octahedral Au core surface or **1** that does not bind covalently to the Au surface. The lifetime of the short-lived, major component [0.13 ns (93%)] is three times as long as that of Au/1 ($n = 11$) (0.040 ns). These results unambiguously exemplify that the quenching of the porphyrin excited singlet state by the gold nanoparticle via energy transfer is much suppressed relative to the energy transfer quenching by the bulk Au(111) surface [146, 148].

We also prepared multiporphyrin-modified metal nanoparticles AuNP-1 with different chain lengths between the porphyrin and the gold nanoparticle to examine the spacer effects ($n = 3,5,7,11$) on the structure and photophysical properties (Fig. 2.6) [148]. The TEM data revealed that the size of gold nanoparticle is not susceptible to the chain length of the spacer even in the case of the large porphyrin moiety. The TEM image of AuNP-1 ($n = 3$) exhibited hexagonal packing of AuNP-1 ($n = 3$) in which the edge-to-edge separation distance between the gold core (3.6 nm) is 2 times as large as the thickness of the porphyrin-monolayer. Such hexagonal packing of AuNP-1 ($n = 3$) can be ascribed to the densely packed, rigid structure of the porphyrin moieties near the gold nanoparticle because of the short methylene spacer [148]. Although no similar hexagonal packing was seen for the other porphyrin-modified gold nanoparticles with a longer spacer ($n = 5,7,11$), the separation distances between the gold core in the TEM images are largely similar irrespective of the chain length of the spacer. Considering that the spacer is splayed outward from the highly curved outermost surface of gold nanoparticles, void space between the porphyrins increases with increasing the chain length of the spacer. This allows the porphyrin moieties to be interdigitated each other to leave the separation distance similar. The fluorescence lifetime was decreased slightly with decreasing the spacer length, which is in accordance with the energy transfer quenching trend in the 2D porphyrin SAMs [104]. Plots of $\ln k_q$ vs d (edge-to-edge distance) yield the same β value (damping factor) of 0.1 ± 0.01 Å^{-1} for AuNP-1 ($n = 3,5,7,11$) and Au/1 ($n = 3,5,7,11$) (Fig. 2.8) [148]. The slower energy transfer rate of the 3D surface than that of the 2D surface may originate from the less gold atoms ($\sim$10^2) on the 3-D surface of the nanoparticles involved in energy transfer in comparison with those on the 2-D surface of the bulk flat electrode. The β value in this study is remarkably small relative to those for conventional energy transfer systems ($0.3 \sim 1.7$ Å^{-1}) [156–160]. The small β value suggests that the alky chain is not fully extended as the chain length increases or that surface plasmons play an important role in the fluorescence quenching, since energy transfer from the exited fluorophore to metal surface is known to be enhanced by surface plasmons and the energy transfer to surface plasmons is a slowly varying function of distance [148]. However, the exact mechanism of fluorescence quenching remains to be clarified.

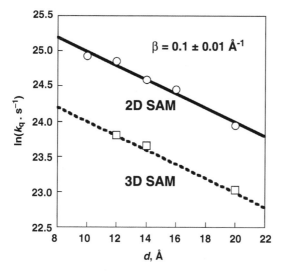

Fig. 2.8. Distance dependence of quenching rate constant for 3D porphyrin SAMs (*dashed line with white squares*) and 2-D porphyrin SAMs (*solid line with white circles*). The plots of $\ln(k_q)$ vs d gave *straight lines* with the slope (-0.1 ± 0.01) according to $k_q = k_0 \exp(-\beta d)$

A variety of porphyrin monolayer-protected metal nanoparticles MNP-1 ($n = 11$) were prepared to examine the effects of metal (M = Au, Ag, Au–Ag alloy, Pd and Pt) and size (i.e., 1~3 nm (M = Au)) on the structures and photophysical properties (Fig. 2.6) [147]. The quenching rate constants of the porphyrin excited singlet state by the surfaces of mono-metal nanoparticles and gold particles with a different diameter are virtually the same. In contrast, the quenching rate constant of the gold-silver alloy nanoparticles is smaller by a factor of $1/2$ than that of the corresponding mono-metal particles (i.e., Au or Ag). This reveals that interaction between the surface of the gold-silver alloy and the porphyrin excited singlet state is reduced considerably in comparison with the mono-metal systems. Accordingly, porphyrin-modified metal nanoparticles are potential candidates as novel artificial photosynthetic materials and photocatalysts [147].

Given that the porphyrin-modified metal nanoparticles possess the high light-harvesting properties together with the suppression effect of energy transfer quenching by the metal surface, we envisaged that they would exhibit efficient energy transfer from the zinc porphyrin excited singlet state to the free base porphyrin when mixed SAMs of zinc porphyrins and free base porphyrins are formed on the gold nanoparticle. Preliminary experiments on the mixed system, however, did not show any clear evidence for the energy transfer process. No energy transfer behavior is rationalized by the relatively large separation distance and the unfavorable parallel orientation between the porphyrins.

However, such morphology is highly favorable to incorporate a guest molecule (i.e., acceptor) between the porphyrins, exhibiting photocatalytic and photovoltaic function. With these in mind, the photocatalytic properties of the porphyrin monolayer-protected gold nanoparticles with different chain lengths were investigated. The photocatalytic reduction of hexyl viologen (HV^{2+}) by 1-benzyl-1,4-dihydronicotinamide (BNAH) was compared with that of the reference porphyrin without the metal nanoparticle **8** (Fig. 2.9) [149]. Both porphyrin monolayer-protected gold nanoparticles and **8** act as efficient photocatalysts for the uphill reduction of HV^{2+} by BNAH to produce 1-benzylnicotinamidinium ion (BNA^+) and hexyl viologen radical cation ($HV^{\bullet+}$) in benzonitrile. In the case of the porphyrin monolayer-protected gold nanoparticle the quantum yield reached a maximum value with an extremely low concentration of HV^{2+}, which is larger than the corresponding value of the reference system using **8**. The dependence of quantum yields on concentrations of BNAH and HV^{2+} as well as the time-resolved single-photon counting fluorescence and transient absorption spectroscopic results indicated that the photoinduced electron transfer from the triplet excited state of **8** to HV^{2+} initiates the photocatalytic reduction of HV^{2+} by BNAH,

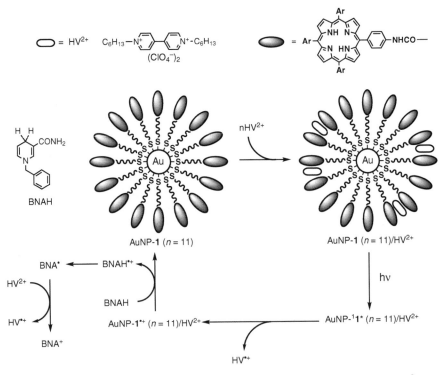

Fig. 2.9. Mechanism of AuNP-1 ($n = 11$)-photocatalyzed reduction of HV^{2+} by BNAH

but that the photoinduced electron transfer from the singlet excited state of porphyrin monolayer-protected gold nanoparticles to HV^{2+}, which forms supramolecular complex with them, is responsible for the photocatalytic reaction [149]. The intersystem crossing from the porphyrin singlet excited state to the triplet excited state is much suppressed by the quenching of the porphyrin excited singlet state via energy transfer to the gold surface of the 3D porphyrin-modified gold nanoparticles. However, the 3D architectures of porphyrin-modified gold nanoparticles with a suitable cleft between the porphyrin moieties allow us to interact HV^{2+} with them, resulting in fast electron transfer from the singlet excited state of porphyrin to HV^{2+} on porphyrin-modified gold nanoparticles [149]. Considering that $HV^{\bullet+}$ can generate H_2 in an acidic aqueous solution using Pt catalyst [161], the present system is fascinating as a photocatalyst for producing H_2, which is expected to play an important role in hydrogen society.

Successful construction of the photocatalytic system using multiporphyrin-modified gold nanoparticles and hexylviologen acceptor has encouraged us to design novel organic solar cells prepared by the bottom-up organization of porphyrin (donor) and fullerene (acceptor) with gold nanoparticles as nanoscaffolds on nanostructured semiconducting electrodes (Fig. 2.10) [150–154]. First, porphyrin disulfide dimers 1 or porphyrin alkanethiols 4 ($n = 5,11,15$) [104] were three-dimensionally organized onto a gold nanoparticle with a diameter of ~2 nm to give multiporphyrin-modified gold nanoparticles AuNP-1 ($n = 5$, 11,15) with well-defined size (~10 nm) and spherical shape (first organization) [146, 148]. These nanoparticles bear flexible host space between the porphyrins for guest molecules (i.e., C_{60}). Although there is equilibrium between the uncomplex and complex states in toluene (second organization), adding poor solvent (i.e., acetonitrile) into the toluene solution triggers the cluster formation in the mixed solvent by $\pi - \pi$ interaction between the porphyrin and C_{60} and the lyophobic interaction between the mixed solvent and the complex. Namely, the nanoparticles AuNP-1 can be grown into larger clusters [denoted as $(AuNP\text{-}1+C_{60})_m$] with a size of ~100 nm in the mixed solvent by incorporating C_{60} molecules between the porphyrin moieties (third organization). Finally, electrophoretic deposition method is applied to the composite clusters in the mixed solvent to give a nanostructured SnO_2 electrode modified with the clusters (denoted as $SnO_2/(AuNP\text{-}1+C_{60})_m/NaI+I_2/Pt$), as shown in Fig. 2.10 (fourth organization). Under application of a dc electric field ($100-500$ V), the clusters in the mixed solvent become negatively charged and are deposited on the SnO_2 electrode as they are driven towards the positively charged electrode surface. The IPCE value of the $SnO_2/(AuNP\text{-}1+C_{60})_m/NaI+I_2/Pt$ device (up to 54% ($n = 15$)) was increased with increasing the chain length ($n = 5,11,15$) between the porphyrin and the gold nanoparticle [150, 151]. The long methylene spacer between the porphyrin and the gold nanoparticle allowed suitable space for C_{60} molecules to accommodate them between the neighboring porphyrin rings effectively compared to the nanoparticles with the short methylene spacer, leading to efficient pho-

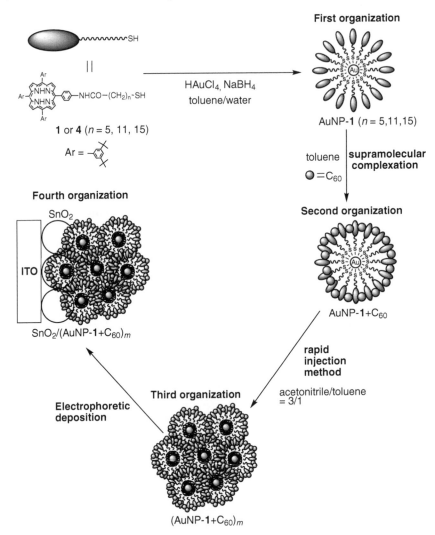

Fig. 2.10. Schematic view of bottom-up organization of porphyrin and fullerene by using gold nanoparticles as nanoscaffolds on a nanostructured SnO_2 electrode for dye-sensitized bulk heteojunction solar cells

tocurrent generation. On the other hand, further increase of the spacer length between the porphyrin and the gold nanoparticle resulted in a substantial decrease of the IPCE value [153]. Additionally, replacement of C_{60} with C_{70} or freebase porphyrin with zinc porphyrin led to a decrease of the photoelectrochemical response [151]. The preference may be explained by the difference in the complexation abilities between the porphyrin and fullerene molecules as well as in the electron or hole hopping efficiency in the composite clusters.

The $SnO_2/(AuNP-1$ $(n = 15)+C_{60})_m/NaI+I_2/Pt$ device had a short circuit current (J_{SC}) of 1.0 mA cm^{-2}, an open circuit voltage (V_{OC}) of 0.38 V, a fill factor (ff) of 0.43, and a power conversion efficiency (η) of 1.5% at a moderate input power (W_{IN}) of 11.2 mW cm^{-2}. The J-V characteristic of the $SnO_2/(AuNP-1$ $(n = 15)+C_{60})_m/NaI+I_2/Pt$ device was also remarkably enhanced by a factor of 45 in comparison with the $SnO_2/(TBPP+C_{60})_m$ device [151]. These results evidently illustrate that the large improvement of the photoelectrochemical properties arises from three-dimensional interdigitated structure of the porphyrin-C_{60} molecules on the SnO_2 electrode, which facilitates the injection of the separated electrons into the CB.

Photocurrent generation is initiated by photoinduced electron transfer from the porphyrin excited singlet state ($^1H_2P^*/H_2P^{\bullet+} = -0.7$ V vs NHE) to C_{60} ($C_{60}/C_{60}^{\bullet-} = -0.2$ V vs NHE) in the porphyrin-C_{60} complex (Fig. 2.11). The reduced C_{60} transfers electrons to the CB of SnO_2 nanocrystallites ($E_{CB} = 0$ V vs NHE) by electron hopping through the large excess of C_{60} molecules, to produce the current in the circuit. The regeneration of $H_2P^{\bullet+}$ ($H_2P/H_2P^{\bullet+} = 1.2$ V vs NHE) is achieved by the iodide/triiodide couple ($I^-/I_3^- = 0.5$ V vs NHE) present in the electrolyte system [151]. Our novel organic solar cells (i.e., dye-sensitized bulk heterojunction (DSBHJ) solar cell) possess both the dye-sensitized and bulk heterojunction characteristics [20–23, 25]. Namely, the device structure is similar to that of dye-sensitized solar cells, but donor-acceptor multilayers are deposited on the top surface of a nanostructured semiconducting electrode. Therefore, initial charge separation takes place at the blend interface of the donor-acceptor, which is typical characteristic of bulk heterojunction solar cells. Nevertheless, DSBHJ solar cells and dye-sensitized solar cells are alike in subsequent processes. It

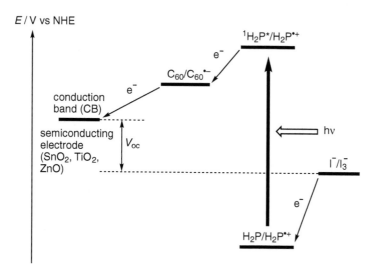

Fig. 2.11. Schematic illustration of photocurrent generation in DSBHJ solar cells

is noteworthy that the composite film reveals the donor-acceptor multilayer structure on a semiconducting electrode, which presents a striking contrast to the monolayer structure of adsorbed dyes on semiconducting electrodes of dye-sensitized solar cells. The sequential electron transfer from the donor excited state to the CB of the semiconducting electrode via the acceptor may inhibit charge recombination between electrons in the CB and donor radical cation or $I_3{}^-$ in the electrolyte owing to the presence of the acceptor, resulting in improvement of the cell performance. Therefore, the photovoltaic properties of DSBHJ solar cells can be modulated by altering device structure including the electrode (SnO_2, TiO_2, ZnO) as well as the donor-acceptor multilayers.

2.3.2 Self-assembled Monolayers of Porphyrins on Semiconducting Nanoparticles

Porphyrin alkanethiols have been successfully assembled on metal nanoparticles using sulfur-metal linkage. The multiporphyrin-modified metal nanoparticles MNP-1 have found to act as light-harvesting materials [146–148], photocatalysts [149] and organic solar cells [150–154]. In the case of metal nanoparticles, however, the relatively fast energy transfer quenching (40–260 ps) of the porphyrin excited singlet state by the metal nanoparticles has still precluded the further improvement of such systems [146–148]. Thus, replacement of the metal core by other nanoparticles is a challenge for exploring novel artificial photosynthetic materials. Along this line there are two promising candidates as nanoparticles: photochemically inactive nanoparticles and semiconducting nanoparticles exhibiting light-harvesting properties. In the former case, we have already reported silica nano- or micro-particles covalently modified with multiporphyrins [162, 163]. The porphyrin excited singlet state is not quenched by the silica nanoparticle. Thus, the photocurrent generation efficiency of the silica nanoparticle-based photoelectrochemical device is significantly higher than that of metal nanoparticle-based corresponding device under the same conditions [163]. The drawback of the system is no light-harvesting properties of the silica nanoparticles that considerably reduce the total light-harvesting efficiency in the multiporphyrin-modified silica nanoparticles. Therefore, nanoparticles, which exhibit efficient light-harvesting, subsequent energy transfer to the immobilized porphyrins on the nanoparticles, and no quenching of the resulting porphyrin excited singlet state by the nanoparticles, are ideal as nanoscaffolds for the construction of efficient solar energy conversion system. Taking into account the requirement, luminescent semiconducting nanoparticles (i.e., CdS, CdSe, CdTe) [164] are potential nanoscaffolds owing to their broad absorption, narrow luminescence, and high photostability. Luminescent semiconducting nanoparticles have been employed as sensitizers in imaging analysis and for an increasing range of applications in biomedicine [165–167]. Such research often deals with energy transfer from luminescent semiconducting nanoparticles to dye molecules in imaging [168–172], photodynamic therapy [173–175], and drug delivery [176]. However, the detailed characterization of dye-luminescent semiconducting nanoparticle composites has been limited because of the complex structure arising from the weak interaction between dye and luminescent semiconducting nanoparticle. For instance, Zenkevich et al. have reported the formation of nanoassemblies consisting of CdSe/ZnS core/shell semiconducting nanoparticles and pyridyl-substituted porphyrin molecules, which were investigated by using UV-visible absorption and steady-state fluorescence spectroscopies and fluorescence life-time measurements [177]. The weak complexation results from the coordination bonding of the pyridyl nitrogen with the ZnS shell of the CdSe/ZnS nanoparticles. The emission quenching of the CdSe/ZnS nanoparticles by the pyridyl-substituted porphyrins was explained partially by energy transfer from the CdSe/ZnS nanoparticles to the porphyrins. In accordance with the fact, only a limited number of vacant binding sites and the weak complexation capability of the CdSe/ZnS nanoparticles for the porphyrins led to less than one porphyrin attached to the ZnS shell of the single CdSe/ZnS nanoparticle. As such, the photophysics of semiconducting nanoparticles covalently modified with chromophores (i.e., porphyrin) has not been fully understood [178–181].

We designed CdSe nanoparticles modified with multiporphyrins [182], as illustrated in Fig. 2.12. CdSe nanoparticles were chosen as a nanoscaffold for organizing porphyrins because of the light-absorbing capability in the UV-visible region, the large band gap (E_g) relative to that of freebase porphyrin (H_2P: 1.9 eV) [146–148], and the relatively facile modification by chromophore [183, 184]. In such a case we can anticipate photoinduced energy transfer from the CdSe nanoparticle to the H_2P in CdSe-1 (Fig. 2.13). Thus, both porphyrin and CdSe nanoparticle are expected to absorb UV-visible light, leading to the eventual production of the porphyrin excited singlet state in CdSe-1. This is in sharp contrast with the multiporphyrin-modified metal or silica nanoparticles where only the porphyrins absorb the UV-visible light [146–148, 163]. More importantly, the porphyrin excited singlet state would not be quenched by the CdSe semiconducting nanoparticle via energy transfer. However, there is possibility of the occurrence of photoinduced electron transfer between the porphyrin and CdSe nanoparticle. The photoinduced electron transfer process depends on the relationship between the CB and valence band (VB) of the CdSe nanoparticle vs the first oxidation and reduction potentials of the porphyrin and the corresponding excited states of the porphyrin. Since it is difficult to anticipate the actual levels of the CB and VB of CdSe nanoparticles, such a study will provide basic and valuable information on the design of chromophore-modified luminescent semiconducting nanoparticles toward efficient solar energy conversion.

The multiporphyrin-modified CdSe nanoparticles CdSe-1 ($n = 11$) were obtained by place-exchange reactions of hexadecylamine-thiophenol-modified CdSe nanoparticles (CdSe-ref) with porphyrin disulfide dimer **1** ($n = 11$) or porphyrin alkanethiol **4** ($n = 11$) in toluene [182]. The number of porphyrin molecules (N) on the surface of single CdSe nanoparticle increased with in-

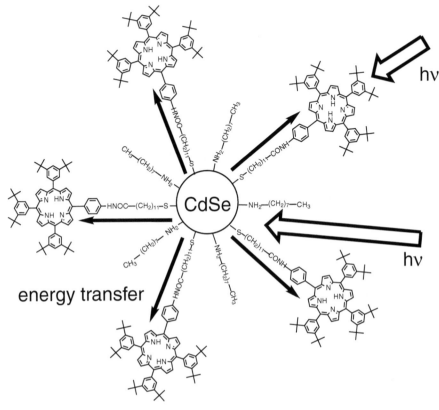

Fig. 2.12. Schematic view of light-harvesting and energy transfer in CdSe-1 ($n = 11$)

creasing the reaction time to reach a saturated maximum of $N = 21$. The structures of CdSe-ref and CdSe-1 were characterized by various spectroscopic methods and surface and elemental analyses [182]. Both of the porphyrins and CdSe nanoparticle in the multiporphyrin-modified CdSe nanoparticle were found to absorb the UV-visible light. The steady-state emission and time-resolved emission lifetime measurements revealed energy transfer from the CdSe excited state to the porphyrins in the multiporphyrin-modified CdSe nanoparticles, as depicted in Fig. 2.14 [182]. The energy transfer efficiency of CdSe-1 ($N = 3.4$) is moderate (33%), whereas the value of CdSe-1 ($N = 21$) is estimated to be ~100%, taking into account the correlation between the N value and the emission intensity from the porphyrins. The emission decay of the porphyrins in CdSe-4 exhibited a single-exponential with a lifetime of 12.3 ns, which is in good agreement with the value of the porphyrin reference **8** (12.3 ns). This exemplifies no quenching of the $^{1}H_2P^{*}$ by the CdSe core, as anticipated from the experimentally determined energy diagram (Fig. 2.13), together with no self-quenching of the $^{1}H_2P^{*}$. These unique prop-

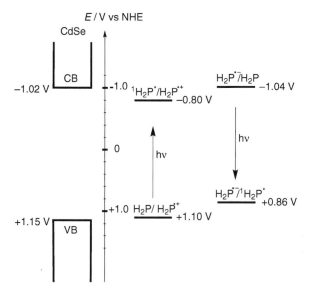

Fig. 2.13. Energy level diagram of CdSe-1 ($n = 11$). The experimental uncertainty to estimate the energy levels is $\pm$ 0.02 V

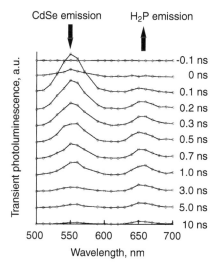

Fig. 2.14. Time-resolved emission spectra of CdSe-1 ($n = 11$, $N = 3.4$) in toluene at $\lambda_{ex} = 375$ nm

erties are in sharp contrast with those in multiporphyrin-modified metal and silica nanoparticles [146–148, 163]. Multiporphyrin-semiconducting nanoparticle composites may be combined with suitable acceptors to develop organic solar cells and photocatalysts, thereby being highly promising as novel artificial photosynthetic materials.

2.4 Molecular Nanostructures

2.4.1 Porphyrin J-Aggregates

Natural chlorophyll aggregates in purple bacteria and chlorosomes have strong transition dipole moments stemming from the alignment of the head-to-tail direction [35–42]. Thus, J-aggregates of synthetic porphyrins are highly promising as light-harvesting models to examine the structure-function relationship. However, the porphyrin J-aggregates have been rather limited to tetrakis(4-sulfonatophenyl)porphyrin (H_2TSPP) and its derivatives in acidic solutions [185–197], protonated tetraphenylporphyrins at the liquid-liquid [198, 199] or gas-liquid interface [199, 200], cationic tetraphenylporphyrins [201], dendritic porphyrins [202], and amphiphilic porphyrins [203]. As such, substituent effects of H_2TSPP on the structures and photophysical properties of J-aggregates have yet to be investigated in detail.

Substituent effects of porphyrin on the structures and photophysical properties of the J-aggregates of protonated 5-(4-alkoxyphenyl)-10,15,20-tris(4-sulfonatophenyl)porphyrin **9** ($n = 1,8,18$) were examined systematically (Fig. 2.15) [204]. The electrostatic interaction between the positively charged

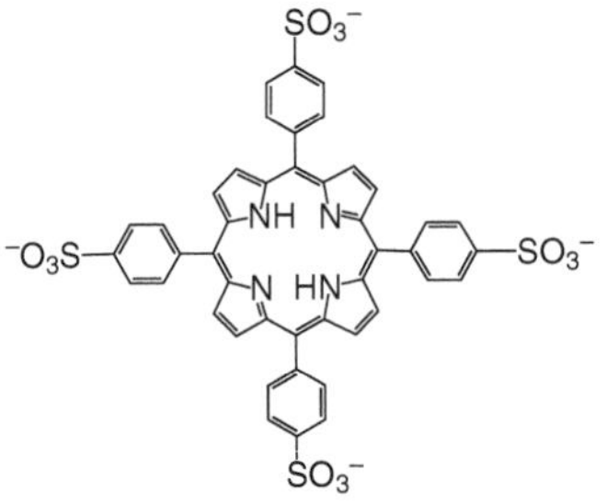

H_2TSPP

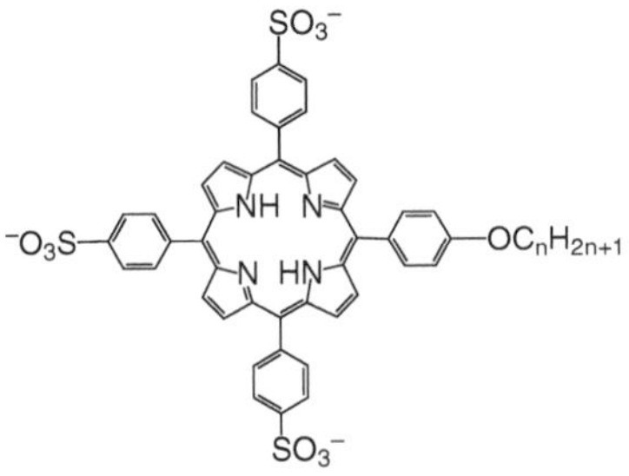

9: n= 1 (OCH$_3$)

n= 8 (OC$_8$H$_{17}$)

n=18 (OC$_{18}$H$_{37}$)

Fig. 2.15. Molecular structures of amphiphilic porphyrins H_2TSPP and **9** ($n = 1,8,18$)

protonated core and negatively charged sulfonato groups is the driving force for the association of the protonated porphyrin monomers to J-aggregates under acidic conditions. Thus, the minimum requirement is the presence of two sulfonato groups at the *para*-positions of the *meso*-phenyl groups in 5,15 positions of the porphyrin ring to attain the slipped head-to-tail structure. To stabilize the J-aggregates, one sulfonate group is replaced by hydrophobic group (i.e., alkoxy group), whereas the three sulfonate groups remain intact. The length of alkyl moiety in the alkoxy group **9** ($n = 1,8,18$) would affect the interaction between the porphyrins under acidic conditions, making it possible to control the structures and photophysical properties of the porphyrin J-aggregates (Fig. 2.15).

Selective formation of the porphyrin J-aggregate was attained for protonated **9** ($n = 8$) [204]. The J-aggregate of the protonated **9** ($n = 8$) displayed the most red-shifted and intense bands, suggesting the highly-ordered architecture. The atomic force microscopy (AFM) image of the J-aggregate from the protonated **9** ($n = 8$) exhibited striking contrast to those from protonated **9** ($n = 1,18$). Regular leaf-like structures (length = 260 nm, width = 60 nm, and height = 4.4 nm) are seen for the protonated **9** ($n = 8$), which largely matches the size (170 nm) determined by the dynamic light-scattering measurements. The cryo-transmission electron microscopy image revealed thin string-like structure with a thickness of 4.9 nm, which is in good accordance with the height value (4.4 nm) determined from the AFM measurements. A bilayer structure was proposed to explain the unique porphyrin J-aggregate in which the hydrophobic alkoxyl groups facing inside the bilayer are interdigitated each other, whereas the hydrophilic porphyrin moieties are exposed outside, as schematically illustrated in Fig. 2.16 [204]. The lifetimes of the

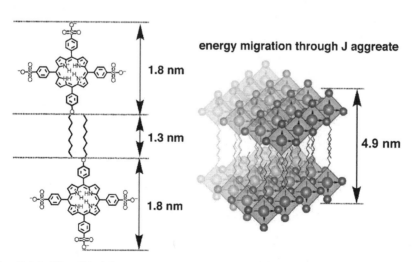

Fig. 2.16. Plausible bilayer structure for the porphyrin J-aggregate of protonated **9** ($n = 8$) in which the alkoxy chains are interdigitated each other, exhibiting a layer thickness of 4.9 nm

J-aggregates for the protonated **9** ($n = 1,8,18$) are much shorter than that for H$_2$TSPP (τ= 350 ps). Furthermore, the lifetime becomes shorter with increasing the length of alkoxy group (τ= 2.3 ps for $n = 1$, τ= 1.7 ps for $n = 8$, $\tau = 1.4$ ps for $n = 18$). Fast energy migration and efficient quenching by defect site in the J-aggregates were suggested to rationalize the short lifetimes of the excited J-aggregates [204].

2.4.2 Conjugated Polymer-Carbon Nanotube Composites

Carbon nanotubes (CNTs) are current targets of general interest for their unique electronic, thermal, mechanical, and optical properties, particularly in connection with exploiting their properties into composites for molecular electronics [205–207]. However, the lack of their solubility in solvents results in a marked impediment toward harnessing their applications. Supramolecular functionalization of CNTs is a potential approach to overcome this problem, because supramolecular interaction does not disrupt unique properties of these composites [208–211]. Surfactants and hydrophilic polymers are known to exfoliate CNT bundles to disperse CNTs in aqueous solutions, but they are still insufficient for the solubilization of CNTs in organic solvents, which are more suitable for the fabrication of CNT composites into molecular devices.

Utilization of $\pi - \pi$ interaction is a promising methodology for dispersing CNTs in organic solvents [208–211]. Various π electron-rich compounds including pyrene, porphyrin, and π-conjugated polymers have been employed to interact with CNTs forming supramolecular composites. Specifically, conjugated polymers have been the subjects of extensive research as active materials for use in light-emitting diodes and photovoltaic devices [212]. Thus, the nanocomposites of highly exfoliated CNTs and conjugated polymers are appealing candidates exhibiting unique photophysical properties in molecular devices. There have been intensive researches on the composites of CNTs and poly(m–phenylenevinylene)-co-(2,5-dioctyloxy-p-phenylenevinylene) [209, 213, 214]. Furthermore, photophysics of the composites of poly(p-phenylenevinylene) (PPV) or polythiophene derivatives and CNTs have been studied in solutions and films [215–219]. Nevertheless, the photophysical properties including energy transfer or electron transfer process between conjugated polymers and CNTs have not been fully elucidated. For instance, only the emission quenching from the π-conjugated polymers in the composites but no emission from the CNTs due to the energy transfer, have yet been observed.

To demonstrate energy transfer from π-conjugated polymers to CNTs unambiguously, it is essential to transform the bundle to isolated individual CNT in solvents by wrapping it with novel π-conjugated polymers. A novel conjugated polymer, poly[(p-phenylene-1,2-vinylene)-co-(p-phenylene-1,1-vinylidene)] (coPPV), was prepared by the Heck coupling reaction to examine specific interactions with single-walled carbon nanotubes (SWNTs), as schematically illustrated in Fig. 2.17 [220]. The coPPV has the structural defect in the main chain of all-$trans$ phenylene-1,2-vinylenes caused

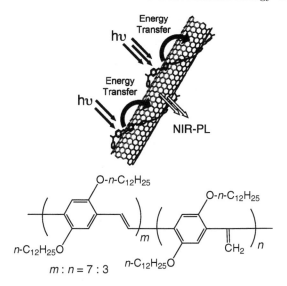

Fig. 2.17. Schematic view of energy transfer from co-PPV to SWNT in the composite, followed by near-infrared photoluminescence from the SWNT

by 1,1-vinylidene moieties. It is expected that, as a result of the introduction of defect sites by 1,1-vinylidene units, the backbone structure of the copolymer can be fitted to the curvature of SWNTs more efficiently than the corresponding regular homopolymer, PPV, yielding individual SWNTs wrapped with the *co*PPV. The absorption and fluorescence properties of *co*PPV (vide infra) disclose that the band gap energy (2.5 eV) exceeds that of SWNTs. Accordingly, *co*PPV is anticipated to exfoliate CNTs more efficiently than the corresponding PPV to debundle SWNTs into individual SWNTs, allowing us to detect the energy transfer process from the excited *co*PPV to the SWNTs in the composites (Fig. 2.17).

The absorption peaks associated with transitions between the symmetric van Hove singularities in the nanotube density of states are evident for the visible-near infrared (vis-NIR) absorption spectrum of the redispersed *co*PPV-SWNT nanocomposites in THF [220]. The results demonstrate that *co*PPV acts as an efficient dispersing agent of SWNTs, since the sharpness of these peaks is widely considered to be a measure of the level of exfoliation of SWNT bundles. The AFM image of the composites revealed the SWNTs with a dramatic decrease in the size of bundles relative to SWNTs without dispersants, implying that the involvement of supramolecular interaction with *co*PPV for the debundling of SWNTs. The composite solution of *co*PPV-SWNTs exhibited a set of emissions in the contour plot (Fig. 2.18). The intensity of (10,5) is much stronger than those of the other peaks, suggesting that (10,5) SWNT is exfoliated to be isolated from bundles more efficiently than SWNTs with other chiral indices. It should be emphasized here that the contour plot of *co*PPV-SWNTs discloses enhancement of emission at an excitation wavelength (λ_{em})

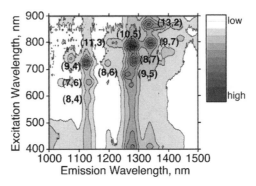

Fig. 2.18. Contour plot of photoluminescence spectra for *co*PPV-SWNTs in THF as a function of excitation and emission wavelengths

of 400–500 nm. The excitation wavelength does not match the electronic absorption peaks of E_{22} (580–900 nm) and E_{33} (<410 nm) of SWNTs included in HiPco samples. Therefore, direct excitation of SWNTs is negligible under the experimental conditions. This emission can be accounted for by the initial excitation due to the $\pi - \pi^*$ transition of *co*PPV, followed by energy transfer from the excited *co*PPV to SWNTs in the composites. The energy transfer process was further substantiated by the excitation spectra of *co*PPV-SWNTs at $\lambda_{em} = 1125$ nm, which are attributed mainly to (9,4) and partially to (7,6) and (8,4) SWNTs [220]. Although there is a report on the enhancement of NIR emission intensity of SWNTs by the energy transfer from an organic molecule encapsulated in SWNTs [221], this is the first observation, to our best knowledge, of the enhancement of emission intensity by interaction with dispersing agents. Our strategy enables the polymer bound to the outside of SWNTs in organic solvents to act as a light-harvesting antenna for SWNTs. The observed characteristic features will be utilized for further exploration of SWNTs as light-emitting and photovoltaic devices.

2.5 Conclusions and Outlook

Synthetic multiporphyrin arrays have been found to be artificial mimics of light-harvesting systems. We have successfully prepared 2D porphyrin arrays on gold and ITO electrodes as well as 3D porphyrin assemblies on metal and semiconducting nanoparticles. In particular, novel multiporphyrin-modified metal nanoparticles revealed improved light-harvesting properties as well as suppression of undesirable energy transfer quenching of the porphyrin excited singlet state by the metal surface. Since multiporphyrin modified metal nanoparticles have flexible molecular recognition clefts between the porphyrins, they can be combined with acceptors (viologens or fullerenes) to exhibit photocatalytic and photovoltaic properties. Specifically, multiporphyrin-modified gold nanoparticles have been assembled with fullerenes in a bottom-up manner to make large and uniform composite clusters on nanostructured semiconductor electrodes, leading to a moderate power conversion efficiency up to 1–2%. Molecular nanostructures including porphyrin J-aggregates and conjugated polymer-SWNT composites have disclosed unique energy transfer behavior based on the well-controlled nanostructures.

Well-defined molecule-based nanoarchitectures exhibiting energy transfer will open the door to nanoscience and nanotechnology, which stimulates a variety of fields including chemistry, biology, physics, and electronics to develop new scientific and technological principles and concepts.

Acknowledgment

The authors are deeply indebted to the work of all collaborators and co-workers whose names are listed in the references (in particular, Prof. P. V. Kamat, Prof. S. Fukuzumi, Prof. H. Lemmetyinen, Prof. N. V. Tkachenko, Prof. S. Isoda, Prof. O. Ito, Prof. D. Kim). H. I. thanks Grand-in-Aids (No. 19350068 to H.I.) and WPI Initiative, MEXT, Japan, for financial support.

References

1. M. R. Wasielewski, Chem. Rev. **92**, 435 (1992)
2. D. Gust, T. A. Moore, A. L. Moore, Acc. Chem. Res. **26**, 198 (1993)
3. D. Gust, T. A. Moore, A. L. Moore, Acc. Chem. Res. **34**, 40 (2001)
4. A. Harriman, J.-P. Sauvage, Chem. Soc. Rev. **26**, 41 (1996)
5. M.-J. Blanco, M. C. Jimenez, J.-C. Chambron, V. Heitz, M. Linke, J.-P. Sauvage, Chem. Soc. Rev. **28**, 293 (1999)
6. V. Balzani, A. Juris, M. Venturi, S. Campagna, S. Serroni, Chem. Rev. **96**, 759 (1996)
7. V. Balzani (ed.), *Electron Transfer in Chemistry* (Wiley-VCH, Weinheim, 2001)
8. N. Armaroli, Photochem. Photobiol. Sci. **2**, 73 (2003)
9. J. Jortner, M. Ratner (eds.) *Molecular Electronics* (Blackwell, London, 1997)
10. M. N. Paddon-Row, Acc. Chem. Res. **27**, 18 (1994)
11. J. W. Verhoeven, Adv. Chem. Phys. **106**, 603 (1999)
12. K. Maruyama, A. Osuka, N. Mataga, Pure Appl. Chem. **66**, 867 (1994)
13. A. Osuka, N. Mataga, T. Okada, Pure Appl. Chem. **69**, 797 (1997)
14. L. Sun, L. Hammarstrom, B. Akermark, S. Styring, Chem. Soc. Rev. **30**, 36 (2001)
15. D. Holten, D. F. Bocian, J. S. Lindsey, Acc. Chem. Res. **35**, 57 (2002)
16. H. Imahori, Y. Sakata, Adv. Mater. **9**, 537 (1997)
17. H. Imahori, Y. Sakata, Eur. J. Org. Chem. 2445 (1999)
18. H. Imahori, Y. Mori, Y. Matano, J. Photochem. Photobiol. C **4**, 51 (2003)
19. H. Imahori, Org. Biomol. Chem. **2**, 1425 (2004)
20. H. Imahori, S. Fukuzumi, Adv. Funct. Mater. **14**, 525 (2004)
21. H. Imahori, J. Phys. Chem. B **108**, 6130 (2004)

22. H. Imahori, J. Mater. Chem. **17**, 31 (2007)
23. H. Imahori, Bull. Chem. Soc. Jpn. **80**, 621 (2007)
24. T. Umeyama, H. Imahori, Energy Environ. Sci. **1**, 120 (2008)
25. H. Imahori, T. Umeyama, J. Phys. Chem. C **113**, 9029 (2009)
26. H. Imahori, T. Umeyama, S. Ito, Acc. Chem. Res. **42**, (2009), in press
27. D. M. Guldi, Chem. Commun. **321** (2000)
28. D. M. Guldi, M. Prato, Acc. Chem. Res. **33**, 695 (2000)
29. D. M. Guldi, Chem. Soc. Rev. **31**, 22 (2002)
30. M. E. El-Khouly, O. Ito, P. M. Smith, F. D'Souza, J. Photochem. Photobiol. C **5**, 79 (2004)
31. L. Sanchez, N. Martin, D. M. Guldi, Chem. Soc. Rev. **34**, 31 (2005)
32. J. H. Alstrum-Acevedo, M. K. Brennaman, T. Meyer, J. Inorg. Chem. **44**, 6802 (2005)
33. M. R. Wasielewski, J. Org. Chem. **71**, 5051 (2006)
34. S. Fukuzumi, Phys. Chem. Chem. Phys. **10**, 2283 (2008)
35. J. Deisenhofer, J. R. Norris (eds.), *The Photosynthetic Reaction Center* (Academic Press, San Diego, 1993)
36. R. E. Blankenship, M. T. Madigan, C. E. Bauer (eds.), *Anoxygenic Photosynthetic Bacteria* (Kluwer Academic Publishing, Dordrecht, 1995)
37. G. McDermott, S. M. Prince, A. A. Freer, A. M. Hawthornthwaite-Lawless, M. Z. Papiz, R. J. Cogdell, N. W. Isaacs, Nature **374**, 517 (1995)
38. A. W. Roszak, T. D. Howard, J. Southall, A. T. Gardiner, C. J. Law, N. W. Isaacs, R. J. Cogdell, Science **302**, 1969 (2003)
39. C. J. Law, A. W. Roszak, J. Southall, A. T. Gardiner, N. W. Isaacs, N. W. Cogdell, Mol. Membrane Biol. **21**, 183 (2004)
40. J. Deisenhofer, O. Epp, K. Miki, R. Huber, H. Michel, Nature **318**, 618 (1985)
41. G. Fritzsch, J. Koepke, R. Diem, A. Kuglstatter, L. Baciou, Acta Crystallogr. **D58**, 1660 (2002)
42. J. M. Olson, Photochem. Photobiol. **67**, 61 (1998)
43. A. Zouni, H.-T. Witt, J. Kern, P. Fromme, N. Kraub, W. Saenger, P. Orth, Nature **409**, 739 (2001)
44. A. Zouni, H.-T. Witt, J. Kern, P. Fromme, N. Kraub, W. Saenger, P. Orth, Nature **411**, 909 (2001)
45. A. Ben-Shem, F. Frolow, N. Nelson, Nature **426**, 630 (2003)
46. S. Prathapan, T. E. Johnson, J. S. Lindsey, J. Am. Chem. Soc. **115**, 7519 (1993)
47. R. W. Wagner, T. E. Johnson, J. S. Lindsey, J. Am. Chem. Soc. **118**, 11166 (1996)
48. J. Li, A. Ambroise, S. I. Yang, J. R. Diers, J. Seth, C. R. Wack, D. F. Bocian, D. Holten, J. S. Lindsey, J. Am. Chem. Soc. **121**, 8927 (1999)
49. R. K. Lammi, A. Ambroise, T. Balasubramanian, R. W. Wagner, D. F. Bocian, D. Holten, J. S. Lindsey, J. Am. Chem. Soc. **122**, 7579 (2000)
50. D. Kuciauskas, P. A. Liddell, S. Lin, T. E. Johnson, S. J. Weghorn, J. S. Lindsey, A. L. Moore, T. A. Moore, D. Gust, J. Am. Chem. Soc. **121**, 8604 (1999)
51. K. Maruyama, A. Osuka, Pure Appl. Chem. **62**, 1511 (1999)
52. T. Nagata, A. Osuka, K. Maruyama, J. Am. Chem. Soc. **112**, 3054 (1990)
53. A. Osuka, N. Tanabe, S. Nakajima, K. Maruyama, J. Chem. Soc., Perkin Trans. **2**, 199 (1996)

54. O. Mongin, A. Schuwey, M. A. Vallot, A. Gossauer, Tetrahedron Lett. **121**, 8927 (1999)
55. N. Solladié, M. Gross, J.-P. Gisselbrecht, C. Sooambar, Chem. Commun. 2206 (2001)
56. R. V. Slone, J. T. Hupp, Inorg. Chem. **36**, 5422 (1997)
57. A. Prodi, M. T. Indelli, C. J. Kleverlaan, F. Scandola, E. Alessio, T. Gianferrara, L. G. Marzilli, Chem. Eur. J. **5**, 2668 (1999)
58. M. J. Crossley, P. L. Burn, J. Chem. Soc., Chem. Commun. 1569 (1999)
59. M. J. Crossley, L. J. Govenlock, J. K. Prashar, J. Chem. Soc., Chem. Commun. 2379 (1995)
60. D. L. Officer, A. K. Burrell, D. C. W. Reid, Chem. Commun. 1657 (1996)
61. M. G. H. Vicente, M. T. Cancilla, C. B. Lebrilla, K. M. Smith, Chem. Commun. 1261 (1998)
62. M. G. H. Vincente, L. Jaquinod, K. M. Smith, Chem. Commun. 1771 (1998)
63. A. K. Burrell, D. L. Officer, P. G. Plieger, D. C. W. Reid, Chem. Rev. **101**, 2751 (2001)
64. V. S.-Y. Lin, S. G. DiMagno, M. J. Therien, Science **264**, 1105 (1994)
65. V. S.-Y. Lin, M. J. Therien, Chem. Eur. J. **1**, 645 (1995)
66. J. T. Fletcher, M. J. Therien, J. Am. Chem. Soc. **122**, 12393 (2000)
67. K. Susumu, M. J. Therien, J. Am. Chem. Soc. **124**, 8550 (2002)
68. H. L. Anderson, S. J. Martin, D. D. C. Bradley, Angew. Chem. Int. Ed. Engl. **33**, 655 (1994)
69. H. L. Anderson, Inorg. Chem. **33**, 972 (1994)
70. H. L. Anderson, S. Anderson, J. K. M. Sanders, J. Chem. Soc., Perkin Trans. 1, 2231 (1995)
71. T. E. O. Screen, J. R. G. Thorne, R. G. Denning, D. G. Bucknall, H. L. Anderson, J. Am. Chem. Soc. **124**, 9712 (2002)
72. C. C. Mak, N. Bampos, J. K. M. Sanders, Angew. Chem. Int. Ed. **37**, 3020 (1998)
73. S. Anderson, H. L. Anderson, J. K. M. Sanders, Angew. Chem. Int. Ed. Engl. **31**, 907 (1992)
74. S. Anderson, H. L. Anderson, A. Bashall, M. McPartlin, J. K. M. Sanders, Angew. Chem. Int. Ed. Engl. **31**, 907 (1992)
75. R. A. Hyacock, A. Yartsev, U. Michelson, V. Sundström, C. A. Hunter, Angew. Chem. Int. Ed. **39**, 3616 (2000)
76. Y. Kuroda, K. Sugou, K. Sasaki, J. Am. Chem. Soc. **122**, 7833 (2000)
77. K. Sugou, K. Sasaki, K. Kitayama, T. Iwaki, Y. Kuroda, J. Am. Chem. Soc. **124**, 1182 (2002)
78. N. Aratani, A. Osuka, Bull. Chem. Soc. Jpn. **74**, 1361 (2001)
79. N. Aratani, A. Osuka, H. S. Cho, D. Kim, J. Photochem. Photobiol. C **3**, 25 (2002)
80. D. Kim, A. Osuka, J. Phys. Chem. A **107**, 8791 (2003)
81. D. Kim, A. Osuka, Acc. Chem. Res. **37**, 735 (2004)
82. A. Osuka, H. Shimidzu, Angew. Chem. Int. Ed. **36**, 735 (1997)
83. N. Aratani, A. Osuka, Y. H. Kim, D. H. Jeong, D. Kim, Angew. Chem. Int. Ed. **39**, 1458 (2000)
84. Y. H. Kim, D. H. Jeong, D. Kim, S. C. Jeoung, H. S. Cho, S. K. Kim, N. Aratani, A. Osuka, J. Am. Chem. Soc. **123**, 76 (2001)

85. N. Aratani, H. S. Cho, T. K. Ahn, S. Cho, D. Kim, H. Sumi, A. Osuka, J. Am. Chem. Soc. **125**, 9668 (2003)
86. A. Nakano, A. Osuka, I. Yamazaki, T. Yamazaki, Y. Nishimura, Angew. Chem. Int. Ed. **38**, 1350 (1999)
87. A. Nakano, T. Yamazaki, Y. Nishimura, I. Yamazaki, A. Osuka, Chem. Eur. J. **6**, 3254 (2000)
88. A. Nakano, A. Osuka, T. Yamazaki, Y. Nishimura, S. Akimoto, I. Yamazaki, A. Itaya, M. Murakami, H. Miyasaka, Chem. Eur. J. **7**, 3134 (2001)
89. Y. Nakamura, N. Aratani, A. Osuka, Coord. Chem. Rev. 831 (2007)
90. K. Sugiura, H. Tanaka, T. Matsumoto, T. Kawai, Y. Sakata, Chem. Lett. 1193 (1999)
91. M. R. Benites, T. E. Johnson, S. Weghorn, L. Yu, P. D. Rao, J. R. Diers, S. I. Yang, C. Kirmaier, D. F. Bocian, D. Holten, J. S. Lindsey, J. Mater. Chem. **12**, 65 (2002)
92. M.-S. Choi, T. Aida, T. Yamazaki, I. Yamazaki, Angew. Chem. Int. Ed. **40**, 3194 (2001)
93. M.-S. Choi, T. Aida, T. Yamazaki, I. Yamazaki, Chem. Eur. J. **8**, 2667 (2002)
94. M.-S. Choi, T. Aida, H. Luo, Y. Araki, O. Ito, Angew. Chem. Int. Ed. **42**, 4060 (2003)
95. M.-S. Choi, T. Yamazaki, I. Yamazaki, T. Aida, Angew. Chem. Int. Ed. **43**, 150 (2004)
96. T. Kato, N. Maruo, H. Akisada, T. Arai, N. Nishino, Chem. Lett. **96**, 890 (2000)
97. E. K. L. Yeow, K. P. Ghiggino, J. N. H. Reek, M. J. Crossley, A. W. Bosman, A. P. H. J. Schenning, E. W. Meijer, J. Phys. Chem. B **104**, 2596 (2000)
98. T. Akiyama, H. Imahori, Y. Sakata, Chem. Lett. **104**, 1447 (1994)
99. T. Akiyama, H. Imahori, A. Ajavakom, Y. Sakata, Chem. Lett. 907 (1996)
100. H. Imahori, H. Norieda, S. Ozawa, K. Ushida, H. Yamada, T. Azuma, K. Tamaki, Y. Sakata, Langmuir **14**, 5335 (1998)
101. H. Imahori, S. Ozawa, K. Ushida, M. Takahashi, T. Azuma, A. Ajavakom, T. Akiyama, M. Hasegawa, S. Taniguchi, T. Okada, Y. Sakata, Bull. Chem. Soc. Jpn. **72**, 485 (1999)
102. H. Imahori, Y. Nishimura, H. Norieda, H. Karita, I. Yamazaki, Y. Sakata, S. Fukuzumi, Chem. Commun. 661 (2000)
103. H. Imahori, H. Yamada, S. Ozawa, K. Ushida, Y. Sakata, Chem. Commun. 1165 (1999)
104. H. Imahori, H. Norieda, Y. Nishimura, I. Yamazaki, K. Higuchi, N. Kato, T. Motohiro, H. Yamada, K. Tamaki, M. Arimura, Y. Sakata, J. Phys. Chem. B **104**, 1253 (2000)
105. H. Imahori, H. Yamada, Y. Nishimura, I. Yamazaki, Y. Sakata, J. Phys. Chem. B **104**, 2099 (2000)
106. H. Imahori, H. Norieda, H. Yamada, Y. Nishimura I. Yamazaki, Y. Sakata, S. Fukuzumi, J. Am. Chem. Soc. **123**, 100 (2001)
107. H. Imahori, T. Hasobe, H. Yamada, Y. Nishimura, I. Yamazaki, S. Fukuzumi, Langmuir **17**, 4925 (2001)
108. H. Yamada, H. Imahori, S. Fukuzumi, J. Mater. Chem. **12**, 2034 (2002)
109. H. Yamada, H. Imahori, Y. Nishimura, I. Yamazaki, S. Fukuzumi, Chem. Commun. **109**, 1921 (2000)

110. H. Yamada, H. Imahori, Y. Nishimura, I. Yamazaki, S. Fukuzumi, Adv. Mater. **14**, 892 (2002)
111. H. Yamada, H. Imahori, Y. Nishimura, I. Yamazaki, T. K. Ahn, S. K. Kim, D. Kim, S. Fukuzumi, J. Am. Chem. Soc. **125**, 9129 (2003)
112. T. Hasobe, H. Imahori, H. Yamada, T. Sato, K. Ohkubo, S. Fukuzumi, Nano Lett. **3**, 409 (2003)
113. T. Hasobe, H. Imahori, K. Ohokubo, H. Yamada, T. Sato, Y. Nishimura, I. Yamazaki, S. Fukuzumi, J. Porphyrins Phthalocyanines **7**, 296 (2003)
114. H. Imahori, K. Hosomizu, Y. Mori, T. Sato, T. K. Ahn, S. K. Kim, D. Kim, Y. Nishimura, I. Yamazaki, H. Ishii, H. Hotta, Y. Matano, J. Phys. Chem. B **108**, 5018 (2004)
115. H. Imahori, M. Kimura, K. Hosomizu, S. Fukuzumi, J. Photochem. Photobiol. A **166**, 57 (2004)
116. H. Imahori, M. Kimura, K. Hosomizu, T. Sato, T. K. Ahn, S. K. Kim, D. Kim, Y. Nishimura, I. Yamazaki, Y. Araki, O. Ito, S. Fukuzumi, Chem. Eur. J. **10**, 5111 (2004)
117. K. Uosaki, T. Kondo, X.-Q. Zhang, M. Yanagida, J. Am. Chem. Soc. **119**, 8367 (1997)
118. T. Kondo, T. Kanai, K. Iso-o, K. Uosaki, Z. Phys. Chem. **212**, 23 (1999)
119. H. Imahori, M. Arimura, T. Hanada, Y. Nishimura, I. Yamazaki, Y. Sakata, S. Fukuzumi, J. Am. Chem. Soc. **123**, 335 (2001)
120. H. Imahori, S. Fukuzumi, Adv. Mater. **13**, 1197 (2001)
121. H. Tamiaki, T. Miyatake, R. Tanikaga, A. R. Holzwarth, K. Schaffner, Angew. Chem. Int. Ed. Engl. **35**, 772 (1996)
122. R. A. Haycock, C. A. Hunter, D. A. James, U. Michelsen, L. R. Sutton, Org. Lett. **2**, 2435 (2000)
123. K. Ogawa, Y. Kobuke, Angew. Chem. Int. Ed. **39**, 4070 (2000)
124. R. Takahashi, Y. Kobuke, J. Am. Chem. Soc. **125**, 2372 (2003)
125. A. Satake, Y. Kobuke, Tetrahedron **61**, 13 (2005)
126. I. W. Hwang, M. Park, T. K. Ahn, Z. S. Yoon, D. M. Ko, D. Kim, F. Ito, Y. Ishibashi, S. R. Khan, H. Miyasaka, C. Keda, R. Takahashi, K. Ogawa, A. Satake, Y. Kobuke, Chem. Eur. J. **11**, 3753 (2005)
127. A. Satake, Y. Kobuke, Org. Biomol. Chem. **5**, 1679 (2007)
128. P. Ballester, R. M. Gomila, C. A. Hunter, A. S. H. King, L. J. Twyman, Chem. Commun. 38 (2003)
129. C. M. Drain, J.-M. Lehn, J. Chem. Soc., Chem. Commun. 2313 (1994)
130. J. Fan, J. A. Whiteford, B. Olenyuk, M. D. Levin, P. J. Stang, E. B. Fleischer, J. Am. Chem. Soc. **121**, 2741 (1999)
131. R. K. Kumar, S. Balasubramanian, I. Goldberg, Inorg. Chem. **37**, 541 (1998)
132. T. Imamura, K. Fukushima, Coord. Chem. Rev. **198**, 133 (2000)
133. M. Fujitsuka, O. Ito, H. Imahori, K. Yamada, H. Yamada, Y. Sakata, Chem. Lett. 721 (1999)
134. H. Imahori, K. Tamaki, D. M. Guldi, C. Luo, M. Fujitsuka, O. Ito, Y. Sakata, S. Fukuzumi, J. Am. Chem. Soc. **123**, 2607 (2001)
135. D. Hirayama, K. Takimiya, Y. Aso, T. Otsubo, T. Hasobe, H. Yamada, H. Imahori, S. Fukuzumi, Y. Sakata, J. Am. Chem. Soc. **124**, 532 (2002)
136. K.-S. Kim, M.-S. Kang, H. Ma, A. K.-Y. Jen, Chem. Mater. **16**, 5058 (2004)
137. A. N. Shipway, E. Katz, I. Willner, ChemPhysChem **1**, 18 (2000)
138. P. V. Kamat, J. Phys. Chem. B **106**, 7729 (2002)

139. M. Brust, M. Walker, D. Bethell, D. J. Schiffrin, R. Whyman, J. Chem. Soc., Chem. Commun. 801 (1994)
140. A. C. Templeton, W. P. Wuelfing, R. W. Murray, Acc. Chem. Res. **33**, 27 (2000)
141. G. Schmid, *Clusters and Colloids. From Theory to Applications* (VCH, New York, 1994)
142. A. P. Alivisatos, Science **271**, 933 (1996)
143. A. Henglein, Ber. Bunsen-Ges. Phys. Chem. **99**, 903 (1995)
144. M. P. Pileni, New J. Chem. 693 (1998)
145. S. Link, M. A. El-Sayed, J. Phys. Chem. B **103**, 4212 (1999)
146. H. Imahori, M. Arimura, T. Hanada, Y. Nishimura, I. Yamazaki, Y. Sakata, S. Fukuzumi, J. Am. Chem. Soc. **123**, 335 (2001)
147. H. Imahori, Y. Kashiwagi, T. Hanada, Y. Endo, Y. Nishimura, I. Yamazaki, S. Fukuzumi, J. Mater. Chem. **13**, 2890 (2003)
148. H. Imahori, Y. Kashiwagi, Y. Endo, T. Hanada, Y. Nishimura, I. Yamazaki, Y. Araki, O. Ito, S. Fukuzumi, Langmuir **20**, 73 (2004)
149. S. Fukuzumi, Y. Endo, Y. Kashiwagi, Y. Araki, O. Ito, H. Imahori, J. Phys. Chem. B **107**, 11979 (2003)
150. T. Hasobe, H. Imahori, S. Fukuzumi, P. V. Kamat, J. Am. Chem. Soc. **125**, 14962 (2003)
151. T. Hasobe, H. Imahori, P. V. Kamat, T. K. Ahn, S. K. Kim, D. Kim, A. Fujimoto, T. Hirakawa, S. Fukuzumi, J. Am. Chem. Soc. **127**, 1216 (2005)
152. H. Imahori, A. Fujimoto, S. Kang, H. Hotta, K. Yoshida, T. Umeyama, Y. Matano, S. Isoda, Adv. Mater. **17**, 1727 (2005)
153. H. Imahori, A. Fujimoto, S. Kang, H. Hotta, K. Yoshida, T. Umeyama, Y. Matano, S. Isoda, Tetrahedron **62**, 1955 (2006)
154. H. Imahori, A. Fujimoto, S. Kang, H. Hotta, K. Yoshida, T. Umeyama, Y. Matano, S. Isoda, M. Isosomppi, N. V. Tkachenko, H. Lemmetyinen, Chem. Eur. J. **11**, 7265 (2005)
155. T. S. Ahmadi, S. L. Logunov, M. A. El-Sayed, J. Phys. Chem. **100**, 8053 (1996)
156. A. M. Brun, A. Harriman, J. Am. Chem. Soc. **116**, 10383 (1994)
157. H. Oevering, J. W. Verhoeven, M. N. Paddon-Row, E. Cotsaris, N. S. Hush, Chem. Phys. Lett. **143**, 488 (1988)
158. B. Schlicke, P. Belser, L. De Cola, E. Sabbioni, V. Balzani, J. Am. Chem. Soc. **121**, 4207 (1999)
159. F. Barigelletti, L. Flamigni, M. Guardigli, A. Juris, M. Beley, S. Chadorowsky-Kimmes, J.-P. Collin, J.-P. Sauvage, Inorg. Chem. **35**, 136 (1996)
160. G. L. Closs, P. Piotrowiak, J. M. MacInnis, G. R. Fleming, J. Am. Chem. Soc. **110**, 2652 (1988)
161. H. Kotani, K. Ohkubo, Y. Takai, S. Fukuzumi, J. Phys. Chem. B **110**, 24047 (2006)
162. H. Imahori, K. Mitamura, T. Umeyama, K. Hosomizu, Y. Matano, K. Yoshida, S. Isoda, Chem. Commun. 406 (2006)
163. H. Imahori, K. Mitamura, Y. Shibano, T. Umeyama, Y. Matano, K. Yoshida, S. Isoda, Y. Araki, O. Ito, J. Phys. Chem. **110**, 11399 (2006)
164. *Nanoparticles*, ed. by G. Schmid (Wiley-VCH, Weinheim, 2004)
165. A. P. Alivisatos, Science **271**, 933 (1996)
166. E. Klarreich, Nature **413**, 450 (2001)

167. A. P. Alivisatos, Nature Biotech. **22**, 47 (2004)
168. J. K. Jaiswall, H. Mattoussi, J. M. Mauro, S. M. Simon, Nature Biotech. **21**, 47 (2003)
169. A. R. Clapp, I. L. Medintz, J. M. Mauro, B. R. Fisher, M. G. Bawendi, H. Mattoussi, J. Am. Chem. Soc. **126**, 301 (2004)
170. I. Medintz, H. Uyeda, E. Goldman, H. Mattoussi, Nature Mater. **4**, 435 (2005)
171. X. Gao, Y. Cui, R. M. Levenson, L. W. K. Chung, S. Nie, Nature Biotech. **22**, 969 (2004)
172. P. K. Chattopadhyay, D. A. Price, T. F. Harper, M. R. Betts, J. Yu. E. Gostick, S. P. Perfetto, P. Goepfert, R. A. Koup, S. C. De Rosa, M. P. Bruchez, M. Roederer, Nature Med. **12**, 972 (2006)
173. A. C. S. Samia, X. Chen, C. Burda, J. Am. Chem. Soc. **125**, 15736 (2003)
174. R. Bakalova, Z. Ohba, Z. Zhelev, M. Ishikawa, Y. Baba, Nature Biotech. **22**, 1360 (2004)
175. A. R. Clapp, I. L. Medintz, H. Mattousssi, ChemPhysChem **7**, 47 (2005)
176. I. Roy, T. Y. Ohulchanskyy, H. E. Pudavar, E. J. Bergey, A. R. Oseroff, J. Morgan, T. J. Dougherty, P. N. Prasad, J. Am. Chem. Soc. **125**, 7860 (2003)
177. E. Zenkevich, F. Cichos, A. Shulga, E. P. Petrov, T. Blaudeck, C. von Borczyskowski, J. Phys. Chem. B **109**, 8679 (2005)
178. O. Schmelz, A. Mews, T. Basche, A. Hermann, K. Müllen, Langmuir **17**, 2861 (2001)
179. I. Robel, V. Subramanian, M. Kuno, P. V. Kamat, J. Am. Chem. Soc.**128**, 2385 (2006)
180. L. Seeney-Haj-Ichia, B. Basnar, I. Willner, Angew. Chem. Int. Ed. **44**, 78 (2005)
181. S. Dayal, Y. Lou, A. C. S. Samia, J. C. Berlin, M. E. Kenney, C. Burda, J. Am. Chem. Soc. **128**, 13974 (2006)
182. S. Kang, M. Yasuda, H. Miyasaka, H. Hayashi, T. Umeyama, Y. Matano, K. Yoshida, S. Isoda, H. Imahori, ChemSusChem **1**, 254 (2008)
183. S. L. Cumberland, K. M. Hanif, A. Javier, G. A. Khitrov, G. F. Strouse, S. M. Woessner, C. S. Yun, Chem. Mater. **14**, 1576 (2002)
184. M. G. Berrettini, G. Braun, J. G. Hu, G. F. Strouse, J. Am. Chem. Soc. **126**, 7063 (2004)
185. D. L. Akins, H.-R. Zhu, C. Guo, J. Phys. Chem. **98**, 3612 (1994)
186. M. A. Castriciano, A. Romeo, V. Villari, N. Micali, L. M. Scolaro, J. Phys. Chem. B **107**, 8765 (2003)
187. R. Rubires, J.-A. Farrera, J. M. Ribo, Chem. Eur. J. **7**, 436 (2001)
188. C. Escudero, J. Crusats, I. Diez-Pe rez, Z. El-Hachemi, J. M. Ribo, Angew. Chem. Int. Ed. **45**, 8032 (2006)
189. R. Rubires, J. Crusats, Z. El-Hachemi, T. Jaramillo, M. Lopez, E. Valls, J.-A. Farrera, J. M. Ribo, New. J. Chem. 189 (1999)
190. N. Micali, V. Villari, M. A. Castriciano, A. Romeo, L. M. Scolaro, J. Phys. Chem. B **110**, 8289 (2006)
191. N. Micali, F. Mallamace, A. Romeo, R. Purrello, L. M. Scolaro, J. Phys. Chem. B **104**, 5897 (2000)
192. A. D. Schwab, D. E. Smith, C. S. Rich, E. R. Young, W. F. Smith, J. C. de Paula, J. Phys. Chem. B **107**, 11339 (2003)

193. R. Rotomskis, R. Augulis, V. Snitka, R. Valiokas, B. Liedberg, J. Phys. Chem. B **108**, 2833 (2004)
194. Y. Kitahara, Y. Kimura, K. Takazawa, Langmuir **22**, 7600 (2006)
195. O. Ohno, Y. Kaizu, H. Kobayashi, J. Chem. Phys. **99**, 4128 (1993)
196. T. Nagahara, K. Imura, H. Okamoto, Chem. Phys. Lett. **381**, 368 (2003)
197. J.-J. Wu, N. Li, K.-A. Li, F. Liu, J. Phys. Chem. B **112**, 8134 (2008)
198. K. Fujiwara, S. Wada, H. Monjushiro, H. Watarai, Langmuir **22**, 2482 (2006)
199. S. Okada, H. Segawa, J. Am. Chem. Soc. **125**, 2792 (2003)
200. L. Zhang, Q. Lu, M. Liu, J. Phys. Chem. B **107**, 2565 (2003)
201. K. Kano, K. Fukuda, H. Wakami, R. Nishiyabu, R. F. Pasternack, J. Am. Chem. Soc. **122**, 7494 (2000)
202. T. Yamaguchi, T. Kimura, H. Matsuda, T. Aida, Angew. Chem. Int. Ed. **43**, 6350 (2004)
203. G. de Miguel, K. Hosomizu, T. Umeyama, Y. Matano, H. Imahori, M. T. Martin-Romero, L. Camacho, ChemPhysChem **9**, 1511 (2008)
204. K. Hosomizu, M. Oodoi, T. Umeyama, Y. Matano, K. Yoshida, S. Isoda, M. Isosomppi, N. V. Tkachenko, H. Lemmetyinen, H. Imahori, J. Phys. Chem. B **112**, 16517 (2008)
205. R. Andrews, D. Jacques, D. Qian, T. Rantell, Acc. Chem. Res. **35**, 1008 (2002)
206. M. Ouyang, J.-L. Huang, C. M. Lieber, Acc. Chem. Res. **35**, 1018 (2002)
207. M. in het Panhuis, J. Mater. Chem. **16**, 3598 (2006)
208. E. Katz, I. Willner, ChemPhysChem **5**, 1084 (2004)
209. P. Liu, Eur. Polym. J. **41**, 2693 (2005)
210. Y. Lin, S. Taylor, H. Li, K. A. S. Fernando, L. Qu, W. Wang, L. Gu, B. Zhou, Y.-P. Sun, J. Mater. Chem. **14**, 527 (2004)
211. H. Murakami, N. Nakashima, J. Nanosci. Nanotechnol. **6**, 16 (2006)
212. T. A. Skotheim, R. L. Elsenbaumer, J. R. Reynolds, *Handbook of Conducting Polymers 2nd ed.* (Marcel Dekker, New York, 1998)
213. S. A. Curran, P. M. Ajayan, W. J. Blau, D. L. Carroll, J. N. Coleman, A. B. Dalton, A. P. Davey, A. Drury, B. McCarthy, S. Maier, A. Strevens, Adv. Mater. **10**, 1091 (1998)
214. A. Star, J. F. Stoddart, D. Steuerman, M. Diehl, A. Boukai, E. W. Wong, X. Yang, S.-W. Chung, H. Choi, J. R. Heath, Angew. Chem. Int. Ed. **40**, 1721 (2001)
215. D. B. Romero, M. Carrard, W. D. Heer, L. Zuppiroli, Adv. Mater. **8**, 899 (1996)
216. J. Wery, H. Aarab, S. Lefrant, E. Faulques, Phys. Rev. B **67**, 115202 (2003)
217. S. Kazaoui, N. Minami, B. Nalini, Y. Kim, K. Hara, J. Appl. Phys. **98**, 084314 (2005)
218. S. Kazaoui, N. Minami, B. Nalini, Y. Kim, N. Takada, K. Hara, Appl. Phys. Lett. **87**, 211914 (2005)
219. H. Zhao, S. Mazumdar, C.-X. Sheng, M. Tong, Z.V. Vardeny, Phys. Rev. B **73**, 075403 (2006)
220. T. Umeyama, N. Kadota, N. Tezuka, Y. Matano, H. Imahori, Chem. Phys. Lett. **444**, 263 (2007)
221. K. Yanagi, K. Iakoubovskii, S. Kazaoui, N. Minami, Y. Maniwa, Y. Miyata, H. Kataura, Phys. Rev. B **74**, 155420 (2006)

3

Electro-Magneto-Optics in Polarity-Controlled Quantum Structures on ZnO

H. Matsui and H. Tabata

3.1 Introduction

Studies have shown that zinc oxide (ZnO) is a practical candidate for the development of practical devices such as thin film transistors, transparent electrodes, and so on [1, 2]. ZnO has a large exciton energy of 60 meV, which raises the interesting possibility of utilizing its excitonic effects at temperatures higher than 300 K [3]. Optically pumped UV stimulated emissions from ZnO layers have been demonstrated [4, 5]. Furthermore, $Mg_xZn_{1-x}O$ alloys are attracting a great deal of interest since they possess a larger band gap than ZnO [6] and have been utilized for $Mg_xZn_{1-x}O/ZnO$ multiple and single-quantum wells [7, 8]. These structures can form low-dimensional systems and produce interesting quantum phenomena, such as increased excitonic binding energy [9, 10] and two-dimensional (2D) electron transport [11] aspect that contribute to both basic science and practical applications.

A variety of nanostructures in semiconductor materials have been made and investigated until. The number of papers concerning nanostructures of ZnO is increasing yearly. Self-organized techniques provide advantages for nanoscale engineering and have yielded many impressive results. Therefore, surface nanostructures in Si and GaAs have been fabricated using various growth mechanisms. Stranski-Krastanov (S-K) growth on lattice mismatched systems induces three-dimensional (3D) nanodots on 2D wetting layers [12]. Lateral surface nanowires have been fabricated due to a step-faceting mode on vicinal surfaces [13, 14]. These surface nanostructures have been developed for zero-dimensional (0D) quantum dots and one-dimensional (1D) quantum wires [15, 16]. Low-dimensional properties are currently receiving attention as advantages for optoelectronics with ZnO.

In epitaxial growth, lattice mismatch between an epilayer and substrate plays a crucial role. Growth studies concerning ZnO epitaxy have been carried out using c- and a-sapphires [17, 18]. Heteroepitaxial layers have a high dislocation density of $10^9 - 10^{10} cm^{-2}$ due to large mismatches in the lattice structure and in thermal expansion [19]. The use of ZnO substrate not only allows the reduction of the number of lattice defects involved in the epilayers, but also permits the selection of various growth directions without any lattice mismatch, which results in a direct understanding of growth dynamics. The growth polarity in ZnO is the primary factor. Zn (0001) and O (000-1) polarities have isotropic atom arrangements and possess spontaneous polarization along growth directions. On the other hand, the M (10-10)-nonpolar surface has an anisotropic atom structure, and the spontaneous polarization occurs parallel to a surface plane [20]. For example, Zn-polar growth produces atomically flat surfaces due to a layer-by-layer mode [21–23]. Whereas M-nonpolar ZnO layers result in anisotropic morphologies with a nanowires structure based on the step-edge barrier effect [24]. Thus, the difference in growth directions influences the surface state, as well as optical and electrical properties of ZnO layers, which can be made more conspicuous through quantum structures. Quantum structures on various surface morphologies exhibit novel electronic and optical properties because quantized energy levels can be tailored by varying the geometric dimensions.

In the last 10 years, manipulating the spin of an electron rather than its charge has opened fascinating new fields for information processing on diluted magnetic semiconductors (DMS), which has emerged as "Spintronics" [25]. II-V DMS is characterized by s, $p - d$ exchange interaction s, p-d exchange interactions between the localized $3d$ spins and the extended band states, opening new fields both in fundamental research and applications. $Cd_{1-x}Mn_xTe$ compounds have been made a practical as Faraday devices [26]. Recently, much interest has centered on magnetic functionality in ZnO DMS because of its magneto-optical effect and ferromagnetic properties [27, 28]. In particular, many researchers have focused on understanding the origin of ferromagnetism in $Zn_{1-x}Co_xO$ from experimental and theoretical viewpoints. Moreover, $Zn_{1-x}Co_xO$ has functionality as an alloy material and has a higher band gap than ZnO [29], which can be utilized in "band gap engineering" and "spin engineering". Therefore, one of the most exciting studies has dealt with advancements in DMS containing heterostructures. In DMS heterostructures, magnetism at the heterointerface can differ from the magnetism of the corresponding bulk materials [30]. A problem with $Zn_{1-x}Co_xO$ is that excitonic emissions are strongly suppressed when increasing doping content. This fact severely hinders the development of spin-dependent emitter devices that utilize excitonic technology characteristics of ZnO. However, a superlattice with quantum wells geometry has the ability to spatially separate excitons from localized $3d$ spins and also retain excitonic emissions. Recently, we succeeded in fabricating a $Zn_{1-x}Co_xO/ZnO$ superlattice with sharp heterointerfaces using the homoepitaxial technique, based on precise understanding of the alloy parameters, growth modes and magnetic properties of $Zn_{1-x}Co_xO$ [31]. This lays the foundations for quantum spin-photonics with ZnO.

This chapter is organized as follows. In Sect. 3.2, we will first give a description of homoepitaxial growth of Zn-polar ZnO layers and $Mg_xZn_{1-x}O/ZnO$ heteroepitaxy. Fabrication of multiple-quantum wells and their low-dimensional optical properties are discussed. In Sect. 3.3, we focus on self-organized surface nanowires on M-nonpolar ZnO layers, wherein discussions concentrate on growth mechanism and developments for the low-dimensional structure, "quantum wires". Section 3.4 is devoted to discussion of various properties of $Zn_{1-x}Co_xO$ DMS and to the fabrication of the quantum wells geometry. Concluding remarks and future research directions in this field are given in Sect. 3.5.

3.2 Zn-Polarity and Quantum Structures

3.2.1 Surface Character

ZnO has a hexagonal wurtzite structure ($a = 0.3250$ nm, $c = 0.5201$ nm), each Zn^{2+} ion bonded to four O^{2-} ions in a tetrahedral formation, representing a structure that can be described as alternating planes of Zn and O ions stacked along the c-axis. Various surface-sensitive methods have been used to investigate the polar surfaces of ZnO from fundamental and applied points of view. For example, the surface morphology was quite different for opposite polar surfaces when ZnO crystals were chemically etched [32]. Thus, epitaxy in ZnO with varying polarity should show different kinetics and material characteristics. Therefore, it is important to understand the uppermost surface structure and morphology in a Zn-polar surface.

Figure 3.1(a) shows a structural model of the bulk-terminated Zn-polar (0001) surfaces of ZnO. All O atoms on the borders have three nearest neighbors, i.e., only one bond is broken. The Zn-polar surface is unstable due to the existence of non-zero dipole moment perpendicular to the surface, which raises a fundamental question regarding stabilization mechanisms [33]. Figure 3.2 shows the AFM image of the Zn-polar (0001) surface of a hydrothermal ZnO substrate annealed at 1100°C in air. The annealed surface had a double-layer step structure whose step height was about 0.26 nm, which corresponds to the half lattice parameter of the unit cell of the c-axis. Analysis of the annealed surface using reflection high-energy electron diffraction (RHEED) revealed 2D streaks attributed to a (1×1) structure, as shown in the inset of Fig. 3.1(b). This shows that no lattice reconstruction occurred in a direction parallel to the surface. However, there is lattice relaxation due to the polarity that occurs along the c-axis. Coaxial impact-collision ion scattering spectroscopy (CAICISS) is useful for surface analyses since this technique is sensitive to the atomic configuration of layer surfaces [34]. The sample was mounted on a two-axis goniometer with respect to the primary He^+ ion beam of ~2 keV at a repetition rate of 100 kHz. The time-averaged current of the incident ion beam was ~150 pA. The CAICISS time-of-flight spectrum taken by a microchannel plate is composed of peaks corresponding to head-on collisions between incident He^+ ions and target atoms on the surface. The Zn signal intensity measured in an ultra-high vacuum of 10^{-10} Torr is shown

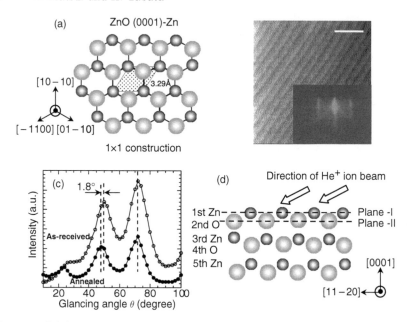

Fig. 3.1. (a) Structural models showing the bulk-terminated Zn-polar (0001) surfaces of ZnO. The surface unit cells are indicated. (b) AFM image of ZnO (0001) surface after annealing at high temperature. (c) Incident angle dependence of the Zn signal intensity when the sample was tilted along the $< 11 - 20 >$ azimuth. (○) and (●) show as-received and annealed Zn-polar surfaces, respectively. (d) Surface models of the Zn-terminated (0001) surface along the [11-20] direction. The *open arrows* indicate the direction of the He$^+$ ion beam

in Fig. 3.1(c), and is consistent with results reported previously [35]. Three peaks of the as-received surface were observed at θ = 23.5, 49.9 and 72.4°. The peaks of the annealed surface then appeared at θ = 23.5, 48.1 and 72.4°. Therefore, the peak angle of 49.9° of the as-received surface was 1.9° lower than that for the annealed surface. Here we define two planes: one is along the [11-20] direction including Zn ions in the uppermost layer (plane-I) as shown in Fig. 3.1(d), and the other is parallel to plane-I consisting of O ions in the second layer (plane-II). The peaks at 23.5° and 72.4° are related to the focusing effect within plane-II, being independent of the uppermost Zn ions, while the middle peaks of 49.9° and 48.1° are ascribed to the uppermost Zn ions (plane-I). To be precise, the middle peak is due to the focusing effect of Zn ions on the first layers to Zn ions on the fifth layers. Thus, the uppermost Zn ions of the annealed surface relaxed toward the inside on the real surface, supported by the peak shift from 49.9° to 48.1°. However, the annealed surface distorts slightly compared to the surface without lattice relaxation (θ = 47.5°), indicating that lattice spacing (Zn–O bond length) is reduced between the uppermost Zn layer and the second O layer. Thus, the Zn-polar surface resulted in no lattice reconstruction in a direction parallel to the surface, while a slight distortion was induced along the c-axis.

3.2.2 Homoepitaxial Growth and Optical Properties

ZnO layers on sapphire usually have O-polarity [19]. Reports have appeared that deal with polarity conversion of ZnO layers on sapphire using buffer layers such as MgO and Cr_2O_3 [36, 37] since Zn-polarity shows a two-dimensional (2D) mode, speculated from the growth of GaN layers with Ga polarity. However, a large lattice mismatch and thermal expansion coefficient between a layer and substrate plays a crucial role in the performance of the Zn-polar layers. Polarity-controlled epilayers can easily be obtained using ZnO substrates and are necessary for the formation of precise device structures such as quantum wells.

Figure 3.2 shows AFM images of 300 nm-thick ZnO layers grown at 420°C. The top surface of the layer grown under an O_2 gas flow of 1.4×10^{-4} mbar was completely covered by highly facetted pit features (Fig. 3.2(a)). The layer grown under an O_2 plasma exposure of 1.4×10^{-5} mbar also exhibited a pitted surface, although the layer appeared to be quite smooth between the pits (Fig. 3.2(b)). An O_2 plasma exposure in the range of 1.4 to 6.0×10^{-4} mbar resulted in few pits with areas possessing a very flat surface (Fig. 3.2(c)

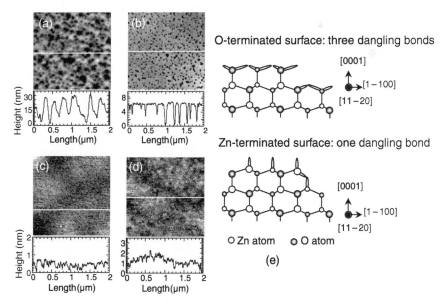

Fig. 3.2. AFM images of ZnO layers grown under an oxygen pressure "$p(O_2)$" of (a) 1.4×10^{-4}, (b) 1.4×10^{-5}, (c) 1.4×10^{-4}, and (d) 6.0×10^{-4} mbar. Oxygen flux was supplied by O_2 gas flow (a) or O_2 plasma exposure (b–d). (e) Schematic surfaces along the Zn-polar direction in the ZnO lattice

and (d)). The surface roughness was about 0.5 nm, a value corresponding to the c-axis length. X-ray diffraction measurements showed line-widths of ω-rocking curves for the (002) and (100) planes were as narrow as 42 and 47 arcsec, respectively.

Figure 3.1(e) shows schematic images of O- and Zn-terminated surfaces along the c-axis. Negatively charged growth surfaces stabilized by oxygen rich conditions is indespensable for 2D mode with Zn-polarity . Each surface atom on O- and Zn-terminated surfaces has one and three dangling bonds, respectively, suggesting that the O sticking coefficient on Zn-terminated surfaces is lower than the Zn sticking coefficient on O-terminated surfaces. The RF plasma source efficiently generates atomic oxygen (O^*), which enhances the O sticking coefficient on the growing surfaces. In contrast, O_2 molecule results in a low surface reaction due to a high binding energy (5.12 eV). The rough surface on the layer grown under O_2 gas flow is ascribed to incomplete O-terminated surfaces. In contrast, O_2 plasma exposure supplies excited O^* atoms to the growing surfaces. O sticking coefficient increased by the excited O^* atoms contributes to the smooth conversion from a Zn-terminated surface to an O-terminated surface. This facilitates formation of negatively charged surfaces and leads to stabilization of the 2D mode. The observed morphological transitions are associated with the variation in the coverage of the layer surface by O atoms.

The variation of exciton peak positions with temperature can be seen in Fig. 3.3(a). The inset of Fig. 3.3(a) shows the photoluminescence (PL) spectrum at 10 K of the Zn-polar ZnO layer at 570°C under O_2 plasma exposure . We identified the 3.377 and 3.384 eV peaks as the free $A - (F_A X)$ and $B - (F_B X)$ excitonic transitions, which were consisitent with values reported in the literature for vapor-phase ZnO [38]. We also observed a narrow peak,

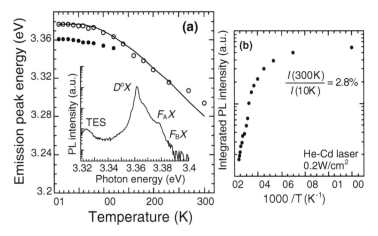

Fig. 3.3. (a) Excitonic transition energies of Zn-polar layers on ZnO substrates as a function of temperature. Inset shows PL spectrum at the band edge. (b) Spectrally integrated PL intensity as a function of T^{-1}

with the strongest peak at 3.362 eV exhibiting a line-width of 2.3 meV, characteristic of a neutral donor-bound exciton (D^0X) peak. Two electron satellites of the D^0X peak were seen at 3.325 eV. The F_AX and D^0X peaks were observed for temperatures ranging from 10 to 120 K. The value of 3.295 eV at 300 K for the F_AX peak was due to the widely accepted room temperature value for the band gap of ZnO (3.37 eV) minus the exciton binding energy of 64 meV [3]. The dependence of F_AX peak intensity on temperature could be fitted using the Bose-Einstein relation with a characteristic temperature (Θ_E) of 335 K, being close to the energy (380 K) for phonon density of state on ZnO. An equivalent internal quantum efficiency (η_{int}^{eq}) throughout this study of near band edge (NBE) emissions at 300 K, which is approximated as the integrated PL intensity divided by that at 10 K $(I_{(300K)}/I_{(10K)})$, directly correlates with τ_{PL} in the ZnO layer $(\eta_{int}^{eq} = 1/(1+\tau_R/\tau_{NR}))$. τ_R and τ_{NR} are the radiative and nonradiative lifetimes, respectively. Integrated PL intensity for NBE emissions as a function of 1/T (Fig. 3.3(b)) show that the value of η_{int}^{eq} was 2.8% at 300 K, which is ten times higher than that of ZnO layers grown on sapphires. In n-type ZnO, Zn vacancy (V_{Zn}) most probably represents defects. Since V_{Zn} produces nonradiative recombination centers in the form of V_{Zn}-defect complexes [39], the suppression of structural defects is enhanced η_{int}^{eq}. Thus, Zn-polar layers on ZnO substrates demonstrates a reduction of structural and point defects.

3.2.3 Mg$_x$Zn$_{1-x}$O/ZnO Heteroepitaxy

The discovery of tunable ZnO band-gap has made the alloy system a promising material for use in the development of optoelectronic devices. Characterization of alloys such as (Mg,Zn)O or (Cd,Zn)O is important from the viewpoint of band-gap engineering and $p-n$ junction. It was found that a Mg$_x$Zn$_{1-x}$O alloy is a suitable material for the barrier layers of ZnO/(Mg,Zn)O super-lattices due to its wider band gap. Since the ionic radius of Mg (0.56 Å) is very close to that of Zn^{2+} (0.60 Å), Mg-rich (Mg,Zn)O alloys with a wurtzite phase have been stably conserved even when a rock salt -tructured MgO is alloyed.

Mg contents doped into a ZnO layer usually depend on the surface polarity, growth technique, and type of substrate. It is known that Ga^{3+} and N^{3-} ions are relatively incorporated on O- and Zn-polarities of ZnO, respectively [40, 41]. Figure 3.4(a) shows the Mg content in Mg$_x$Zn$_{1-x}$O layers as a function of the target Mg content. Under growth conditions in this work, the Mg content in Zn-polar layers was always 1.6 times the content of the ablation targets. This can be attributed to the low vapor pressure of Mg-related species compared to those of Zn. The incorporation efficiency of Mg atoms into the layers is more enhanced for O-polarity. Similar behavior was also observed in Cd$_x$Zn$_{1-x}$O alloys (Fig. 3.4(b)). The amount of Cd atoms in the layers is much lower than that of targets, originating from a difference of vapor pressures between Cd- and Zn-related species. Re-evaporation processes on the growing surfaces strongly dominate the doping efficiency of Mg and Cd atoms

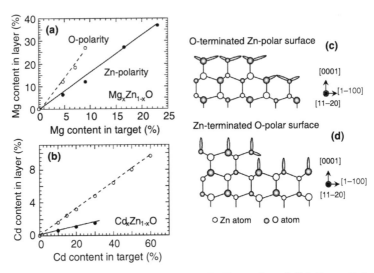

Fig. 3.4. (a) Mg and (b) Cd contents in $Mg_xZn_{1-x}O$ and $Cd_xZn_{1-x}O$ layers as a function of the target Mg and Cd contents, respectively. (b) and (c) show plan-view atomic arrangement of oxygen-terminated Zn-polar and O-polar ZnO surface structures, respectively

in the layers. On the other hand, the polarity dependence is related closely to a sticking coefficient of Zn atoms since this sticking coefficient is higher for Zn-polarity than O-polarity. This results from a number of dangling bonds on O-terminated surface structures of both polarities.

Micro (μ)-photoluminescence and μ-Raman scattering spectroscopy were carried out at room temperature (RT) to study luminescent properties. A fourth-harmonic generation of an yttrium aluminum garnet (YAG) laser at 266 nm was used as excitation source. A spectrum was detected using a 0.85-cm double monochromator. In this measurement, a reflective-type objective lens was used to focus the laser to a diameter of 500 nm on the layer surface (Fig. 3.5). A micro-probe system is an excellent complementary tool for the identification and characterization of the spatial resolution of a microscopic scale region [42].

Figure 3.6(a) shows the c- and a-axis lengths deduced from the (0004) and (11-24) diffraction peaks. When x increased from 0 to 0.37, the c-axis length decreased linearly from around 5.210 and 5.18 Å, while the a-axis length remained at 3.25 Å up to x = 0.27 and then suddenly increased to ~3.27 Å. As shown in Fig. 3.6(b), the unit-cell volume shrinks towards the c-axis direction up to $x = 0.27$, storing the lattice strain and then suddenly expanding toward the a-axis direction by releasing the lattice strain at $x = 0.37$. Consequently, the cell volume comes to resemble that of ZnO substrates. Figure 3.6(c) shows the variation of μ-PL spectra at RT for $Mg_xZn_{1-x}O$ (x = 0–0.37) with layer thicknesses ranging from 200 to 300 nm. Excitonic emissions of all layers

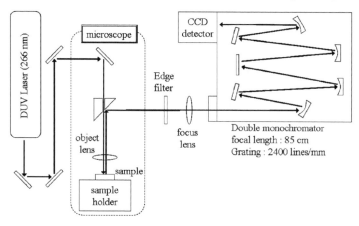

Fig. 3.5. Schematic picture of Optical systems for μ-PL and μ-RRS measurements

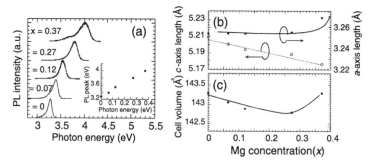

Fig. 3.6. (a) μ-PL spectra of $Mg_xZn_{1-x}O$ layers ($x = 0$–0.37). Inset shows PL peak position as a function of Mg content. (b) Dependence of c- and a-axis lengths on Mg content. (c) Dependence of cell volume on Mg content

systematically shifted from 3.3 to 4.0 eV in a linear manner with increasing x due to the band gap widening. Careful inspection of Fig. 3.6(c) revealed that the layer with $x = 0.37$, which is lattice-strain relaxed (Fig. 3.6(a) and (b)), showed a secondary peak around 3.68 eV. Other layers showed only a single peak. The lattice relaxation promotes phase separation of alloy materials [43].

3.2.4 Stranski-Krastanov Mode and Lateral Composition Modulation

Self-assembled three-dimensional (3D) islands have received a great deal of attention due to the potential fabrication of nanoscale devices using epitaxial growth process. Lattice mismatch between a layer and substrate leads to Stranski-Krastanov (S-K) growth. Adatoms are initially deposited in the form of a 2D pseudomorphic layer, and elastic strain energy that increases with the layer thickness is finally relieved by the formation of 3D dots [44]. S-K growth is of particular interest in strained alloy layers since this growth presents a promising bottom-up technique for the ordered assembly of quantum dots [45]. Due to atomic size differences between cations or anions, alloy layers have stored elastic stress that lead to accelerated growth instability. This hinders coherent growth of alloy layers and results in S-K growth. In particular, force field induced by local alloy fluctuation in strained layers leads to lateral or vertical composition modulation through S-K growth [46].

The surface morphologies of $Mg_{0.37}Zn_{0.63}O$ layers are shown in Fig. 3.6 [47]. The strained layer (thickness "t" = 38 nm) gives a flat (2D) surface, characteristic of Zn-polar growth (Fig. 3.6(a)). At a large thickness, nanodots appeared on the growing surface. When lattice relaxation began (t = 65 nm), nanodots formed spontaneously with a density of 10^{10} cm^2 and a lateral distribution of 46 nm (Fig. 3.6(b)). The partially relaxed layer (t = 100 nm) was uniformly covered by nanodots in a hexagonal shape with an area density of 10^9 cm^2, and the lateral size increased to 153 nm (Fig. 3.6(c)). Finally, the growth scheme in the fully relaxed layer (t = 280 nm) changes to a continuous 3D island growth through the coalescence of nanodots (Fig. 3.6(d)). The lattice parameters as a function of layer thickness are shown in Fig. 3.6(d). When the layer thickness increases, the a-axis length remains at 3.250 Å up to t = 38 nm, compressive stress at the heterointerfaces is estimated to be 0.52%. The a-axis length is then gradually expands, releasing the strain energy up to 3.270 Å in the layer with t = 280 nm, which was fully relaxed. The 2D to 3D transformation in the strained $Mg_{0.37}Zn_{0.63}O$ layer initiated dislocated S-K growth accompanying lattice relaxation at a critical thickness of 38–65 nm. A mechanical equilibrium model by Matthews and Blakeslee [48] gives the critical thickness h_c through the following formulation:

$$h_c = \frac{1}{1+\nu}\frac{1}{4\pi}[\ln\left(\frac{h_c}{b}\right) + 1], \qquad (3.1)$$

where b and ν are the Burgers vector and Poisson ratio, respectively, and f gives the lattice mismatch defined by:

$$f = (a_{MgZnO} - a_{ZnO})/a_{ZnO}, \qquad (3.2)$$

with a_{MgZnO} and a_{ZnO} being the a-axis length of $Mg_{0.37}Zn_{0.63}O$ and ZnO, respectively. For our calculatiions, values of a_{MgZnO} = 3.268 Å, a_{ZnO} = 3.251 Å, f = 0.0052 and ν = 0.271 were used. Furthermore, we set b = 6.137 Å, assuming a Burgers vector of the a/c type for a slip system {1-101}< 11−23 > [49]. The deduced value of h_c = 38 nm was close to the lattice relaxation regions obtained from the experiment results.

Figure 3.7(a) shows the μ-Resonance Raman scattering (μ-RRS) spectra of $Mg_{0.37}Zn_{0.63}O$ layers with different thicknesses. When the thickness increased from t = 38 to 280 nm, the $A_1(LO)$ peak monotonically shifted to a higher frequency (from 620 to 631 cm^{-1}). The asymmetric Γ_a/Γ_b, defined by the

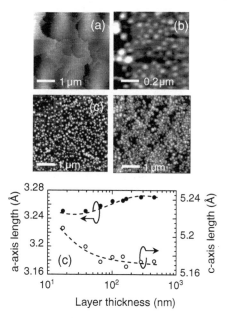

Fig. 3.7. AFM images of $Mg_{0.37}Zn_{0.63}O$ layers with $t = 38$ nm (**a**), 65 nm (**b**), 100 nm (**c**) and 280 nm (**d**). (**e**) Systematic variations of a- and c-axis length in $Mg_{0.37}Zn_{0.63}O$ layers in relation to the layer thickness

ratio of full width at half maximum (FWHM) of the lower-frequency side (Γ_a) to the higher-frequency side (Γ_b), also increased monotonically from 1.97 to 2.49. Figure 3.7(b) compares the spectrum of strained layers ($t = 38$ nm) with different Mg contents, $x = 0.12$ and 0.27. With increasing Mg content, the peak frequency shifted to a higher frequency, expected because of the reduction in reduced mass of the oscillating pairs. Furthermore, the peak broadened with increasing asymmetry Γ_a/Γ_b. These results may indicate that an increase in the Mg content leads to an increase in the randomness of the atomic arrangement and a relaxation of the Raman selection rule. Therefore, the 1st LO phonon of ZnO shows asymmetric broadening, reflecting its phonon DOS (density of states).

When lattice relaxation occurs in the thicker layers, the randomness in atomic arrangement will be more enhanced, resulting in lateral composition separation, as revealed clearly by the results of cathodoluminescence (CL) mapping. CL was measured using a scanning electron microscope equipped with an Oxford mono-CL mirror and grating spectrometer system. Figure 3.8(a) and (b) shows a 10×10 μm^2 topological surface image and the corresponding CL image, respectively of the sample with $t = 280$ nm. The CL and topological surface images were observed simultaneously under measurement conditions of 10 kV and 150 nA. Use of these values made it possible to observe improved CL images. However, the topological surface image resulted

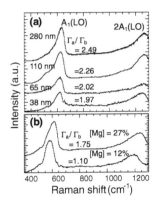

Fig. 3.8. (a) μ-RRS spectra at room temperature of $Mg_{0.37}Zn_{0.63}O$ layers with $t = 38$, 65, 100 and 280 nm. (b) μ-RRS spectra of strained $Mg_{0.12}Zn_{0.88}O$ and $Mg_{0.27}Zn_{0.73}O$ layers

in some degree of smearing since the focusing power of the electron beam slightly diffuses. Compositional fluctuation appears with local CL intensity variations. The CL image is mapped with emission intensity of 3.90 eV using the bright (intense) and dark (weak) scale. There are inhomogeneous regions in the micrometer scale in the CL image. Comparing Figs. 3.7(a) and 3.8(b), the bright and dark regions of the topological image correlates to that of the CL image, as confirmed by the cross-section profiles of X- and Y-lines shown in Fig. 3.8(c) and (d). This suggests a correlation between surface roughening and lateral segregation of Mg atoms. Local composition separation is characterized by a dislocated SK growth of alloy layers. Thus, the origin of surface $Mg_{0.37}Zn_{0.63}O$ nanodots is clearly attributed to a dislocated SK growth.

3.2.5 Multiple Quantum Wells and Excitonic Recombination

For advances concerning epitaxy of ZnO and related alloys, multiple quantum wells (MQWs) are of considerable practical interest. MQWs can provide larger oscillation strength, enhanced excitonic binding energy, and tenability of the operating wavelength due to a quantum confinement effect. The formation of abrupt interfaces between constituent layers is a key issue when fabricating MQWs. In the case of O-polar MQWs on sapphire substrates, the Mg content is strongly limited to 20% because of inhomogeneous heterointerfaces between a well and barrier layers [50]. On the other hand, the use of lattice-matched ScMgAlO₄ imparted a high Mg content of 27% which greatly improved structural quality in O-polar MQWs [8]. As mentioned in the preceding sections, homoepitaxial growth on the Zn-polar ZnO substrate was achieved, and preferential growth in the 2D mode was expected. Furthermore, the 2D growth of MgZnO alloy layers was fully maintained up to Mg content of 37% [43]. The preservation of the 2D growth of MgZnO/ZnO heteroepitaxy is essential for the development of precise device structures, which can provide further enhancement of the degree of freedom in the fabrication of MQWs.

A fabrication scheme for MQWs is as follows. Thickness of barrier layers of $Mg_{0.37}Zn_{0.63}O$ were set to below the critical thickness . After growing a 200 nm-thick ZnO buffer layer on a Zn-polar ZnO substrate, ten periods of a $Mg_{0.37}Zn_{0.63}O$ (13 nm)/ZnO (3.3 nm) structure were grown at 500°C under O_2 plasma exposure , followed by a 10 nm-thick ZnO capping layer (Fig. 3.9(a)). Here, the layer thickness was evaluated by comparing the High-resolution x-ray diffraction (HR-XRD) profile ($2\theta/\omega$ scan) of the (0002) plane, as shown at the top of Fig. 3.9(b), with dynamic kinetic simulation [51] shown in the middle of Fig. 3.9(b). The pronounced pendellosung fringes observed in the HR-XRD profile suggests a high crystal quality for the MQWs since any imperfection or compositional inhomogeneity would decrease the phase coherence and eliminate the pendellosung fringes. The ω-rocking curves of SL-0, SL-1 and SL+1 peaks showed very narrow line-widths [40-50 arcsec]. This suggested small fluctuations in lattice axis orientation in the c-plane. The surface indicated a very flat morphology with a roughness of 0.41 nm. Sharp heterointerfaces between the well and barrier layers were observed by X-ray Transmittance electron microscopy (TEM) analysis as shown in the inset of Fig. 3.10(c).

The optical properties of the MQWs were examined. Figure 3.10 shows the μ-PL spectra at RT observed from the cleaved edge of the (10-10) face [(a) − (d)] and top surface of the MQWs [(e)]. The surface spectrum (e) consisted of a strong broad peak at 3.46 eV and a weak signal around 3.90 eV. The 3.46 eV peak originated from carrier recombination in the ZnO well layers, while the 3.90 eV peak was ascribed to the $Mg_{0.37}Zn_{0.63}O$ layers. This was

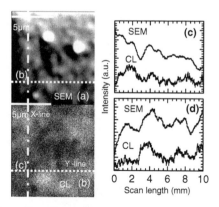

Fig. 3.9. (a) Topological surface image. Note that the two bright spots do not indicate surface morphology, but dust used for a correction of focusing of the electron beam. (b) Monochromatic CL image taken at 3.90 eV. (c) and (d) shows cross-section profiles of X and Y-lines, respectively, indicated by the *white dotted lines* of (a) and (b)

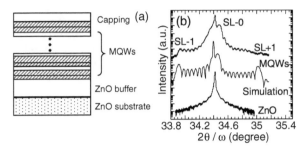

Fig. 3.10. (a) Schematic cross-section of the MQWs. (b) High-resolution $2\theta/\omega$ profiles of the (0002) reflection for MQWs and the ZnO buffer including a theoretical simulation profile

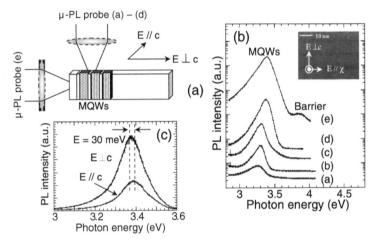

Fig. 3.11. (a) Schematic figure of an excitation method for the MQWs. (b) μ-PL spectra emitted from the surface layer [e] and from the cleaved edge [a]–[d]. Inset shows an X-TEM image of the MQWs. (c) The μ-PL spectra measured with polarization oriented parallel ($E//c$) and perpendicular ($E \perp c$) to the c-axis of the MQWs

supported by the PL spectra of the cleaved cross section. The spatial resolution of the apparatus (~0.5 μm as determined by the laser-beam waist at the sample surface) was smaller than the width of the MQWs constituent layers. Notwithstanding this fact, a systematic peak shift was observed from 3.28 to 3.46 eV (Fig. 3.10(a)–(d)) when moving the laser spot in the cleaved edge cross section from the ZnO substrate to the MQWs side (Fig. 3.10(a)). The 2D effect of the MQWs was confirmed by polarized μ-PL detection from the cleaved edge, as shown in Fig. 3.11(c). The excitation laser with polarization perpendicular to the c-axis of the MQWs ($E \perp c$) yielded a stronger PL signal than that observed with parallel polarization ($E//c$). Here, the polarization degree (P) was calculated as 0.43 by $(I_\perp - I_{//})/(I_\perp + I_{//})$, showing a perfect polarization of the emission, where $I_\perp$ and $I_{//}$ are MQWs emissions under $E \perp c$ and $E//c$ configurations, respectively. The high P value at RT is related to a typical feature of 2D quantum confinement, the exciton can move freely within the MQWs plane but cannot move in a direction perpendicular to this plane [52, 53, 56]. Therefore, the polarization-dependent PL spectra indicate that the excitons are sufficiently confined even at RT. In contrast, the polarized emission of barrier layers was weak due to the 3D nature of excitons. Such high-quality MQWs open up numerous possibilities for UV optoelectronic devices. These favorable properties cannot be attained for MQWs on sapphire substrates due to the lattice mismatch.

3.3 Nonpolarity and Quantum Structures

3.3.1 Nonpolar Growth of M-Face (10-10)

With homoepitaxial ZnO growth, one can select various growth directions without any lattice mismatch at the interface between the film and the substrate. This plays an important role in understanding the growth dynamics concerning epitaxial layers with different growth planes. ZnO layers on polar (0001) and (000-1) surfaces are dominated by 2D growth and spiral step-flow growth modes, respectively, which relate to specific atomic arrangement configurations and number of dangling bonds. In the nonpolar ZnO (10-10) plane, surface Zn and O atoms produce dimer rows running along the [1-210] direction [33]. This produces two types of step edges by polar and nonpolar faces towards the [1-210] and [0001] directions (Fig. 3.15(b)). This type of anisotropic surface morphology has been utilized in scientific studies of heterogeneous catalytic processes involving the absorption of molecular and metallic atoms on nonpolar surfaces [54]. We briefly describe the growth process and morphological evolution of surface nanowires in nonpolar ZnO on the basis of RHEED investigations [55].

At the very beginning of M-nonpolar growth up to 8 nm in thickness, a 2D streak pattern appeared to replace of sharp patterns of ZnO substrates (Fig. 3.12(a) and (b)). This is related to 2D nucleation at the initial growth stage, as evidenced by the smooth layer surface (Fig. 3.12(f)). Continued growth of ZnO changed to a mixed pattern, which relates to the onset of the transition from 2D to 3D modes. This results from the on-set of a self-assembly of anisotropic 3D islands (Fig. 3.12(c) and (g)). Finally, the RHEED pattern showed 3D spots due to an island growth mode that originated from the formation of surface nanowires (Fig. 3.12(d) and (h)). Surface nanowires with high density (10^5 cm^{-1}) that formed on the ZnO layers were homogeneously elongated along the [0001] direction above 5 μm with a few branches.

Due to lattice strains at the heterointerface of layer/substrate, S-K growth naturally induces 3D islands that are surrounded by high-index facets on 2D

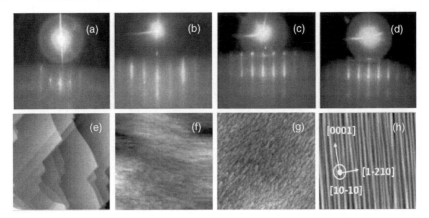

Fig. 3.12. RHEED patterns with the [0001] azimuth of the treated ZnO substrate (**a**) and ZnO layers with a thickness of (**b**) 8, (**c**) 20 and (**d**) 240 nm. AFM top view ($2\times2\,\mu m^2$) of the treated ZnO substrate (**e**) and ZnO layers with different thicknesses (**f**)–(**h**). Layer thicknesses are (**f**) 8, (**g**) 20 and (**h**) 250 nm

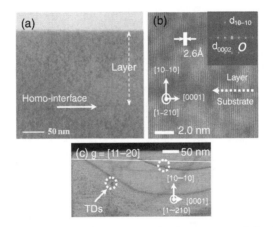

Fig. 3.13. (**a**) Low- and (**b**) high-resolution X-TEM images of the ZnO layer taken with the [11-20] zone axis. Inset shows the RSD obtained by FFT analysis. (**c**) A bright field plan-view TEM image of the ZnO layer with a g = [11-20] excitation under two-beam conditions

wetting layers [12]. This has been observed in InGaAs/GaAs heteroepitaxy [56]. In an effort to examine the crystallinity in greater detail, plane-view and X-TEM observations were conducted to investigate the structural quality of the layer. Figure 3.13 (a) shows a low-resolution X-TEM image with the [11-20] zone axis. Threading dislocations induced by lattice relaxation between the layer and substrate were not observed. The high-resolution X-TEM image of Fig. 3.13(b) reveals a lattice arrangement between a smoothly connected layer and substrate. A 3×3 nm^2 space area selected from the layer region was utilized for fast Fourier transform (FFT) analysis to examine local lattice parameters and yielded a reciprocal space diffractogram (RSD) pattern [inset of Fig. 3.13(b)]. From the RSD pattern, the estimated strains (ε_{yy} and ε_{zz}) at the interface were approximately 0.10% and 0.18% with x, y and z being parallel to the [0001], [10-10] and [11-20] directions, respectively. Figure 3.13(c) shows a bright field plan-view TEM image with g = [11-20] excitation under two-beam conditions. Out-of-plane dislocations, marked by a white open circle, were observed with a density of 3.2×10^7 cm^2, originating from threading dislocations running perpendicular to the layer surface. On the other hand, there were no in-plane dislocations propagating along the [0002] and [11-20] directions from different g vector excitations. These results indicate that the homoepitaxial interface was almost strain free. Thus, the elongated 3D islands that appeared on the 2D layers were formed under coherent homoepitaxy and had no correlation with the S-K growth.

3.3.2 Step-Edge Barrier and Self-organized Nanowires

Similar surface nanowires have been formed by a step-faceting growth mode on vicinal GaAs (331) A and (553) B substrates [13, 14]. The origin of the surface nanowires on non-vicinal ZnO (10-10) substrates differs clearly from that on vicinal GaAs substrates and the S-K mode based on lattice relaxation. On the other hand, it is known that morphological transformation from a 2D surface to anisotropic 3D islands occurs during Si and GaAs (001) homoepitaxy [57, 58]. The Schwoebel barrier effect is considered as an important growth parameter. This barrier mechanism induces growth stability of nucleating anisotropic 3D islands on 2D growing surfaces above a critical thickness. Furthermore, the driving force of anisotropic surface morphology is associated with a difference in surface diffusion and sticking probability along the [110] and [-110] directions of GaAs (001) surfaces. The only research concerning the growth behavior on M-nonpolar ZnO (10-10) surfaces has been reported by Dulub et al. [59]. The anisotropic atomic arrangement of a ZnO (10-10) surface provides a corrugated surface that is related to the anisotropic diffusion coefficient for the growth of Cu on the surface since Cu diffusion was much faster along the [1-210] direction than along the [0001] direction. Cu grew two-dimensionally only up to a certain critical coverage, at which point it began to form 3D islands due to a Schwoebel barrier effect. The growth conversion from 2D to 3D modes of M-nonpolar ZnO layers may be associated with growth instability originating from a Schwoebel barrier effect and the atomic arrangement of surface Zn and O atoms leading to anisotropic surface diffusion.

Figure 3.14(a) and (b) shows low- and high-resolution X-TEM images with the [0001] zone axis, respectively. A cross section of the surface nanowires displayed a triangular configuration with a periodicity of 84 nm. A high-resolution X-TEM image, marked by a white circle, revealed that the side

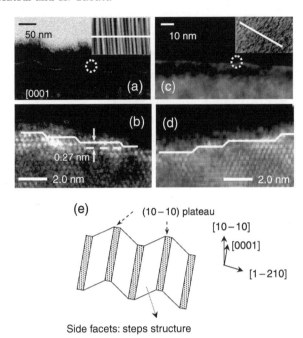

Fig. 3.14. (a) Low- and (b) high-resolution X-TEM images of the ZnO layer with a thickness of 240 nm. (c) Low- and (d) high-resolution X-TEM images of a 20 nm-thick ZnO layer. Insets (a) and (c) represent AFM images of the ZnO layers used for X-TEM observations. (e) Schematic representation of surface nanowires identified from X-TEM images

facets did not consist of high-index facets but instead had a step-like structure with a height of 2.7Å that corresponded to half a unit of the *m*-axis. Side facets of the surface nanowires possessed uniform step spacing ranging from 10 to 20 Å, and were not surrounded by the high-index facets. A large number of surface nanowires showed flat tops with a (10-10) face and were separated laterally by deep grooves, as illustrated schematically in Fig. 3.14(b). A similar structure was also seen in the anisotropic 3D islands on the 20 nm-thick layers, which indicated that the surface nanowires resulted from a coarsening of anisotropic 3D islands formed at the initial growth stage.

A multilayer morphology is determined not only by the transport of atoms within an atomic layer (*intralayer transport*), but also by the transport of atoms between different atomic layers (*interlayer transport*). Thus, evolution of mound shapes is understood in terms of activation of atomic processes along the step edge. Therefore, a sequence of multilayer growth is governed by activation of atomic processes which enable exchange and hopping of atoms between different atomic layers (Fig. 3.15(a)). Schwoebel and Shipsey introduced the schematic potential energy landscape near a step that has become the signature of what is often referred to as the *Ehrlich-Schwoebel barrier*

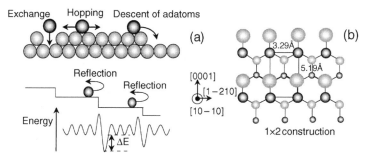

Fig. 3.15. (a) Upper part of the figure shows the descent of adatoms from an island by hopping and exchange. The lower part illustrates the energy landscape for hopping and the definition of ΔE_S. (b) Structural models showing the M-nonpolar (0001) surfaces of ZnO. The surface unit cells are indicated

(ESB) with a barrier energy of ΔE [60]. The mass transport of atoms between different atoms is inhibited by a strong ESB effect, resulting in mound formation. This induces a nucleation of islands on the original surface together with inhibited interlayer transport. Once the islands are formed, atoms arriving on top of the islands will form second layer nuclei, and a third layer will nucleate on top of the second layer. This repetition leads to an increase of surface roughness with increasing layer thickness (Θ), resulting in the formation of mound shapes. Mound formation is often observed on various systems such as semiconductors, metals, and organic materials. A mound structure possesses a small flat plateau at the top and a side facet with constant step spacing, and is separated from other mounds by deep grooves. This structure has been observed on dislocation-free metal homoepitaxial surfaces such as Pt/Pt (111) and Ag/Ag (100) systems and is often referred to as a *wedding cake* [61]. Here, emerging of mound formation under reduced interlayer transport is described by the coarsening $\lambda \sim \Theta^n$, and the surface width $w \sim \Theta^b$. λ and w values are the height-height correlation between the nanowires and the surface roughening, respectively. As seen in Fig. 3.16 (a) and (b), a coarsening exponent was indicated by $n = 0.23$, which was close to the n value of the mound fomration seen after metal homoepitaxy [61]. Furthermore, the high β value of 0.60 was suitable for mound growth based on the ESB. This indicates that the surface nanowires formed during M-nonpolar ZnO homoepitaxy are due to the growth process originating at the ESB. The ESB was also seen for layer growth of O-polar ZnO with a hexagonal island surface [62]. Furthermore, the appearance of an anisotropic morphology is related closely to a difference in surface diffusion and sticking probability as an important parameter. In M-nonpolar ZnO, the stoichiometric surface is auto-compensated since it contains an equal number of Zn and O ions per unit area. Surface Zn and O atoms form dimer rows running along the [1-210] direction, as shown in Fig. 3.15 (b). This produces two types of A and B step edges consisting of stable low-index (1-210) and (0001) planes, respectively. The [1-210] direction

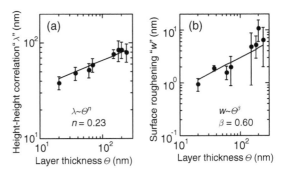

Fig. 3.16. (a) Height-height correlation (λ) and surface roughening (w) as a function of layer thickness

represents an auto-compensated nonpolar surface, while the [0001] direction consists of a polar surface with either Zn or O termination. Thus, the origin of the surface nanowires is based not only on a ESB barrier, but a difference in the surface diffusion and sticking coefficient of atoms between the two types of step edges.

3.3.3 Linearly Polarized Light Emissions

ZnO has attracted great interest for new fields of optical applications. A characteristic of the wurtzite structure is the presence of the polarization-induced electric field along the c-axis. However, the optical quality of a quantum-well structure grown along the c-axis suffers from undesirable spontaneous and piezoelectric polarizations in well layers, which lower quantum efficiency [1]. The use of nonpolar ZnO avoids this problem due to an equal number of cations and anions in the layer surface. Nonpolar ZnO surfaces have in-plane anisotropy of structural, optical, acoustic, and electric properties, which is useful for novel device applications. Recently, the number of investigations concerning nonpolar ZnO heteroepitaxial layers has grown considerably [63, 64], although heteroepitaxial growth involves introducing anisotropic lattice strains that modify the surface morphology and optical transitions. This hinders the elucidation of the intrinsic characteristics of the nonpolar layers and multiple-quantum wells (MQWs). In this session, we discuss polarized PL of M-plane ZnO layers and $Mg_{0.12}Zn_{0.88}O$/ZnO MQWs grown on M-plane ZnO substrates.

Figure 3.17 shows splitting of the valence band (VB) in ZnO under the influence of crystal-field splitting and spin-orbit coupling. The VB of ZnO is composed of p-like orbitals. Spin-orbit coupling leads to a partial lifting of the VB degeneracy, and the former six-fold degenerate VB is split into a four-fold ($j = 3/2$) and two-fold ($j = 1/2$) band. The spin-orbit coupling is negative. The $j = 1/2$ band is at a higher energy than the $j = 3/2$ band. On the other hand, the crystal field in ZnO results in further lifting of VB degeneracy due

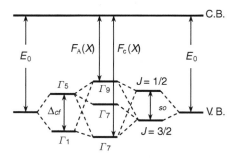

Fig. 3.17. Schematic energy level of band splitting by the crystal-field (Δ_{cf}) and spin-orbit (Δ_{SO}) interactions in a wurtzite structure. $F_A(X)$ and $F_C(X)$ correspond to A and C-excitons, respectively, which are indicated in the middle. In the electronic energy levels proposed by Park et al. [65] and Reynolds et al. [66], the uppermost Γ_9 and Γ_7 levels are interchanged

to the lower symmetry of wurtzite compared to zinc blende. The crystal field causes a splitting of p states into Γ_5 and Γ_1 states. Crystal-field splitting Δ_{cf} and spin-orbit coupling Δ_{SO} together give rise to three two-fold degenerate valence bands. These bands are denoted as A (Γ_9 symmetry), B (Γ_7) and C (Γ_7). These energies can be calculated as follows:

$$E_A(\Gamma_9) - E_B(\Gamma_7) = -\frac{\Delta_{SO} + \Delta_{cf}}{2} + \frac{\sqrt{(\Delta_{SO} + \Delta_{cf})^2 - \frac{8}{3}\Delta_{SO}\Delta_{cf}}}{2} \tag{3.3}$$

$$E_A(\Gamma_9) - E_C(\Gamma_7) = \sqrt{(\Delta_{SO} + \Delta_{cf})^2 - \frac{8}{3}\Delta_{SO}\Delta_{cf}} \tag{3.4}$$

For ZnO, the experimental results were $E_A - E_B = 0.0024$ eV and $E_C = 0.0404$ eV [67]. Solving the above two equations, we obtain Δ_{cf} (0.0391 eV) and Δ_{SO} (-0.0035 eV). A- and B-excitons are referred to as heavy (HH) and light hole (LH) bands, respectively, and the crystal-field split-off hole (CH) was related to the C-exciton. The detection of $E \perp c$ and $E//c$ points to A-exciton (X_A) and C-exciton (X_C), respectively, where E represents the electric field vector [68].

Figure 18(a) shows the $E \perp c$ and $E//c$ components of the normalized PL spectra of strain-free ZnO layers [69]. The peak energies of X_A and X_C were located at 3.377 and 3.419 eV, respectively. These energies coincided with the X_A (3.377 eV) and X_C (3.4215 eV) peaks in ZnO crystals, respectively. The dependence of peak intensities on temperature could be fitted using the Bose-Einstein relation with a characteristic temperature of 315 and 324 K for the X_A and X_C peaks, respectively. The bound exciton ($D^0 X$) peak disappeared at 120 K due to the activation energy of 16 meV. The polarization degree (P) is defined as $(I_\perp - I_{//})/(I_\perp + I_{//})$, where $I_\perp$ and $I_{//}$ are the peak intensities for $E \perp c$ and $E//c$, respectively. Figure 3.18(b) shows the polarization-dependent

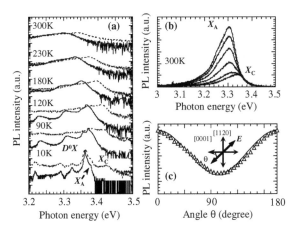

Fig. 3.18. (a) Temperature dependence of PL spectra on strain-free ZnO layers for $E \perp c$ (*solid lines*) and $E//c$ (*dotted lines*). (b) Polarization-dependent PL spectra at 300 K taken in steps of $\Delta\theta = 15°$. (c) PL intensity as a function of polarization angle θ. *Inset* shows a schematic representation of the measurement geometry and sample orientation

PL spectra at 300 K. The layer strongly emitted polarized light. The P value was calculated as 0.49. Significant spectral shifts in PL were detected when altering the polarization angle. This is attributed to a difference in carrier distribution in the VB between the HH and CH levels at 300 K. Figure 3.18(c) shows the dependence of polarization angle on PL intensity. The experimental data (triangle dots) were in agreement with the $\cos(\theta)^2$ fit line (solid) obeyed by Malus' laws.

The polarized PL character in M-nonpolar MQWs were be discussed. ZnO wells are strain-free in the case of pseudomorphically grown MgZnO/ZnO MQWs. The PL spectra for $E \perp c$ and $E//c$ in MQWs with a well thickness (L_W) of 2.8 nm are shown in Fig. 3.19(a). The emission peaks around 3.6 eV correspond to 7 nm-thick $Mg_{0.12}Zn_{0.88}O$ barriers. At 300 K, an energy separation (ΔE) of 37 meV was found between the MQWs emissions of 3.372 eV ($E \perp c$) and 3.409 eV ($E//c$). The emission peak for $E \perp c$ appeared under conditions of $E//c$ below 120 K since the thermal distribution of carriers in the high-energy level for $E//c$ is negligible at 10 K. A polarization degree close to unity was found with a high P of 0.92 at 10 K (Fig. 3.19(c)). In contrast, excited carriers at 300 K were sufficiently distributed in the high-energy level, resulting in a low P of 0.43. Furthermore, ΔE between the emission peaks for $E \perp c$ and $E//c$ was retained at around 40 meV even at 60 K (Fig. 3.19(d)). This ΔE was close to the theoretical ΔE between the X_A and X_C states [70]. For unstrained bulk ZnO, a polarization magnitude of zero and unity in the C-exciton is detected along the normal direction and along the c-axis, respectively (Fig. 3.18(a)). However, the confinement of M-plane MQWs takes place perpendicular to the quantization of the [10-10] direction. This gener-

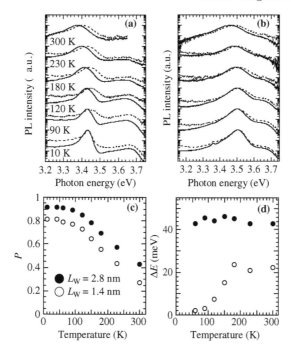

Fig. 3.19. PL spectra under $E \perp c$ (*solid lines*) and $E//c$ (*dotted lines*) on MQWs with $L_W = 2.8$ nm (**a**) and 1.4 nm (**b**). (**c**) and (**d**) show the relationship of temperature with polarization degree (P) and energy separation (ΔE) on MQWs with different L_W

ates weak mutual mixing of the different p orbitals. Thus, a π polarization ($E//c$) component is expected for the A-excitonic state in these MQWs. In the case of M-plane MQWs, it is predicted that a 10% p_z orbital component is involved with the A-excitonic states [71], which is in agreement with the experimentally obtained P value of 0.92. MQWs with a L_W of 1.4 nm showed that the polarized PL spectra of $E \perp c$ and $E//c$ were separated by a small ΔE of 27 meV at 300 K (Fig. 3.19(b)). ΔE decreased with temperature, and then completely disappeared at 60 K. The P value also dropped for all of the temperature regions. These behaviors are due to a large admixture of p_x to p_z orbitals for $E//c$, originating from an inhomogeneous roughening between the well and barrier layers. The interface roughness increased a potential fluctuation of quantized levels in the MQWs, being reflected by the broadened PL lines [72]. This was proven experimentally by the increase in line-width of PL spectra with a narrowing of L_W.

3.3.4 Large Anisotropy of Electron Transport

Self-organization of 1D nanostructures on growth surfaces has attracted much attention, as this phenomenon can form low-dimensional systems such as quantum wires and quantum dots. These low-dimensional super-structures produce interesting quantum phenomena in terms of both scientific and practical applications. Above all, understanding the formation of 1D surface morphology through a bottom-up process represents one of the challenges in crystal growth technology. However, the underlying origin of crystal growth of an anisotropic layer along one direction is still unclear. Recently, the ability to precisely control 2D growth for Zn-polarity in ZnO has blazed a new trail in the fields of quantum physics. Here, a surface roughening that occurs during layer growth provides a simple and efficient way to fabricate low-dimensional surface nanostructures, which can then spatially confine carriers. When spatial undulation occurs at a MgZnO/ZnO heterointerface, electron transport in MQWs shows anisotropic behavior. Previous studies have reported that 2D electron transport of AlGaAs/GaAs MQWs was anisotropically modulated using vicinal GaAs substrates with lateral surface corrugations [73].

A pronounced anisotropy of conductivity was observed in 7-period $Mg_{0.12}Zn_{0.88}O$/ZnO MQWs grown on 200 nm-thick ZnO buffer layers. The barrier thickness was set to be 7 nm, and varying well widths were controlled between 1.4 and 4.0 nm. Surface nanowires elongated along the [0001] direction were retained even after growing the MQWs. Figure 3.20(a) shows an X-TEM image of MQWs with the [0001] zone axis. The layers with a bright contrast represent MgZnO barriers, while the darker layers indicate ZnO wells. The MgZnO layers repeat the surface structure of the underlying ZnO layers,

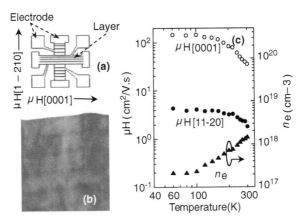

Fig. 3.20. (a) Hall bar pattern used to investigate anisotropic transport. (b) X-TEM image of $Mg_{0.12}Zn_{0.88}O$/ZnO MQWs. (c) Temperature dependence of Hall mobility parallel ($\mu_{H[0001]}$) and perpendicular ($\mu_{H[11-20]}$) to the nanowire arrays and electron concentration (n_e) for MQWs with a $L_W = 3.2$ nm. Electrical properties parallel and perpendicular to the nanowires were measured using a Hall bar configuration with the perpendicular arms of the Hall bar aligned in the [0001] and [1-210] directions (b). The Hall bars were fabricated by Ar ion milling of the samples through a photolithography-defined resist mask

indicating that $Mg_{0.12}Zn_{0.88}O/ZnO$ heterointerfaces are periodically modulated by the surface nanowires.

Figure 3.20(c) shows the temperature-dependent Hall mobility parallel $(\mu_{H[0001]})$ and perpendicular $(\mu_{H[1-210]})$ to the nanowires. The *ex situ* annealed ZnO substrates were treated as a semi-insulating substrate showing electrical resistivity in the order of $10^6 - 10^7$ $\Omega\cdot$cm. $\mu_{H[0001]}$ gradually increased with decreasing temperature and became constant just below 150 K due to a suppression of ionized impurity scattering. The electron concentration of MQWs also saturated below 100 K, suggesting that the entire electric current flows as 2D-like transport through the ZnO wells. In contrast, $\mu_{H[1-210]}$ was much lower and resulted in large anisotropy of electron transport. Figure 3.21(a) shows the ratio of $\mu_{H[0001]}$ and $\mu_{H[1-210]}$ as a function of temperature. The curves correspond to different L_W of 2.2, 2.8 and 4 nm. For MQWs with a L_W of 4 nm, we observed almost no anisotropic behavior. However, the anisotropy of the Hall mobility increased to 52 for MQWs with a L_W of 2.8 nm at low temperature.

The transport properties indicate that electrons can move almost freely along the nanowires, but are blocked from moving perpendicular to the nanowires. We discuss a possible mechanism for this type of activation barrier. The large anisotropy of electron transport disappeared when a flat surface was realized using Zn-polar MQWs, as shown in Fig. 3.21(b). Interface-roughness scattering dominates low-temperature mobility in MQWs [74]. A slight roughness of the heterointerfaces induces a large fluctuation in quantization energy of confined electrons. This acts as a scattering potential barrier for electron motion and reduces mobility. Therefore, electrons may readily undergo frequent scattering in a direction perpendicular to the nanowires by potential barriers produced between nanowires, and, consequently, may become extremely immobile.

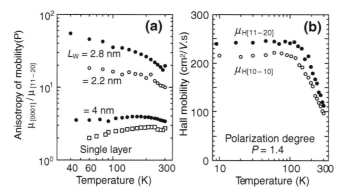

Fig. 3.21. (a) Temperature dependence of anisotropy of mobility (P) for M-nonpolar $Mg_{0.12}Zn_{0.88}O/ZnO$ MQWs with different well thicknesses (L_W) of 2.2, 2.8 and 4 nm. (a) Temperature dependence of Hall mobility $(\mu_{H[11-20]})$ and $(\mu_{H[10-10]})$ for Zn-polar $Mg_{0.27}Zn_{0.73}O/ZnO$ MQWs with a $L_W = 2.4$ nm

In contrast, parallel conductance along the nanowires involves a lower scattering probability than perpendicular transport due to a weak heterointerface modulation. However, the P value for MQWs with a L_W of 2.2 nm decreased with a decrease in $\mu_{H[0001]}$. Inspection of polarized PL spectra showed that the energy fluctuations in the quantum well gradually increased with decreasing L_W (Fig. 3.18). A decreased Hall mobility with a narrowing of L_W has been observed on very thin InAs/GaSb MQWs since energy fluctuations in a quantum well are caused by an increase in interface roughness [75]. It is concluded, therefore, that the large transport anisotropy was obtained through both a quantum size effect and small energy fluctuations in the quantum well, i.e., when L_W was in the vicinity of 3 nm.

3.4 Quantum Well Geometry Based on ZnCoO

3.4.1 Spin and Band Engineering

Diluted ZnCoO magnetic semiconductors (DMS) display s,p-d exchange interactions between the localized magnetic moments of transition-ions and the extended band states that yield a large Zeeman splitting. Although many investigations have been directed towards elucidating the origin of ferromagnetic ordering in ZnCoO [76, 77], detailed studies concerning the modulation of band structure induced by the incorporation of Co ions remain unclear. For non-magnetic $Zn_{1-x}Mg_xO$ alloy layers, excitonic-related optical properties at the band-edge changed systematically with increasing band gap regardless of the synthetic methods employed, while the alloy parameters in ZnCoO tended to be strongly dependent on fabrication technique. CoO is a Mott-Hubbard insulator with a charge transfer gap of 5.0 eV [78], and differs largely from non-magnetic MgO (band gap: 6.2 eV) with respect to electronic structure. It is expected to meet various fatal problems induced by utilizing doping with Co atoms in comparison to Mg atoms. Therefore, a precise understanding of the properties of ZnCoO alloy is essential for the development of spin-optics.

Figure 3.22(a) shows the absorption spectra at 10 K of $Zn_{1-x}Co_xO$ layers ($x = 0$–0.05) grown on C-face sapphires. A free excitonic transition (FX) was clearly observed at 3.387 eV with phonon side bands of 72 meV for un-doped ZnO. The FX peak gradually shifted to a higher energy with increasing Co content, and then the excitonic structure diffused with an obvious band tail around 3.2 eV. Charge transfer (CT) levels from Co^{2+} to $Co^+ + h^+_{VB}$ in $Zn_{1-x}Co_xO$ seem to contribute to the sub-band gap since it hold a photon energy of 3.2 eV [79]. This is further illustrated in detail in the magnetic circular dichroism (MCD) spectra.

Figure 3.22(b) shows the MCD signals obtained at 10 K. An external magnetic field induces weak Zeeman splitting of the semiconductor band structure (Fig. 3.22(c)). In DMS, this splitting can be huge due to the s, $p-d$ exchange interaction. Thus, the effective magnetic field on the sp band electrons is

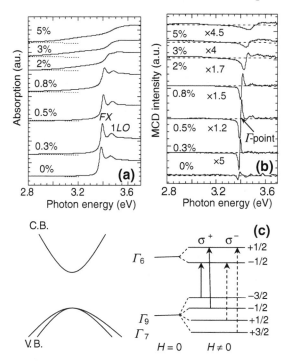

Fig. 3.22. (a) Optical absorption spectra and (b) MCD signals of $Zn_{1-x}Co_xO$ ($x = 0 - 5\%$) layers measured at 10 K. (c) Zeeman splitting and σ^+ and σ^- optical transitions at Γ-points

amplified by the magnetic moment of the transition metal ion through the $s, p - d$ exchange interaction s,p-d exchange interactions. Here, this splitting produces circularly polarized optical anisotropies that are widely known as the MCD effect, which allow a more precise evaluation of the magneto-optical properties of DMS. In general, MCD intensity is expressed by the following relation:

$$MCD = \frac{90}{\pi l} \frac{I^+ - I^-}{I^+ + I^-} \tag{3.5}$$

where l is the thickness of the sample, and I^+ and I^- are the intensities of transmitted light in σ^+ and $\sigma-$ polarizations, respectively. MCD depends on the photon energy and it is usually very strong near a resonant line or band split by the Zeeman effect. The un-doped ZnO layer exhibits a very weak MCD signal at the Γ-point, which originats from excitonic transitions from the Γ_9 and Γ_7 levels of V.B. to C.B. The MCD response in $Zn_{1-x}Co_xO$ ($x \neq 0$) layers was strongly enhanced due to $p - d$ electron coupling between the t_{2g} states of Co^{2+} and the $2p$ state of oxygen. In regions with Co contents above 2%, however, the MCD response is weakened and broadened due to alloy fluctuation. A broader shoulder with negative polarity then appeared at 3.2 eV corresponding to CT levels. The Co 3d (t_{2g}) levels occupied by up-spin are located at a neighbor of the V.B. Thus, the photo-excited CT states would appear in shallow levels near the C.B. As a result, excitonic recombination at the band edge is suppressed due to the appearance of the band tail. $Zn_{1-x}Co_xS$ and $Zn_{1-x}Co_xSe$ have also exhibited circularly polarized CT transitions [80].

The Γ-point in the MCD corresponds to the band gap energy (E_0-edge). The shift in the E_0 edge to a higher energy with increasing Co content is shown in Fig. 3.23. A linear increase in the E_0 edge with increasing Co content (x) was obtained and is expressed by $E(x) = 3.387$ eV $+0.026x$, indicating a band gap widening for $Zn_{1-x}Co_xO$. Here, the reason for the band gap widening can be tentatively explained as follows. An electronic structure of CoO consists of double energy gaps: one represents a charge transfer gap of 5.0 eV from O(2p) to the occupied 3d levels. while the other is a wide gap of 8 eV between the occupied 3d levels and Co(4s) levels, indicating that an optical gap between O(2p) and Co(4s) levels would be very wide [81]. Furdyna et $al.$ stressed the close relationship of s and p electron bands in $A^{II}_{1-x}Mn_xB^{VI}$ systems to the bands of nonmagnetic $A^{II}B^{IV}$ [82]. The band gap widening in $Zn_{1-x}Co_xO$ can be ascribed to hybridization of the s, p band of ZnO with that of CoO. Furthermore, the g-factor from the MCD spectra is qualitatively estimated using the following equation:

$$MCD = -\frac{45}{\pi} \times \Delta E \times \frac{dK}{dE} (\Delta E = g\mu_B H) \tag{3.6}$$

where K is the absorption coefficient, E is the photon energy, μ_B is the Bohr magneton and H is the magnetic field. From Fig. 3.23, the g-factor linearly increases with increasing Co content, and then gradually saturates over 2%. This correlates with an antiferromagnetic (AFM) exchange interaction between two nearest-neighbor Co^{2+} ions and is based on spin coupling viewed

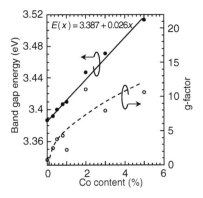

Fig. 3.23. (●) Photon energy derived from the MCD E0 edge at 10 K, which essentially indicates the band gap of $Zn_{1-x}Co_xO$. (○) Dependence of g factor on Co content in $Zn_{1-x}Co_xO$

as an indirect exchange interaction mediated by the anion. This AFM coupling essentially produces a net zero magnetic moment for the pair, resulting in a decrease of total susceptibility. Thus, the probability of obtaining isolated Co^{2+} ions quickly decreases due to nearest-neighbor complexes with increasing Co content, resulting in a decrease of the effective number of isolated spins. Therefore, Co^{2+} ions are almost retained as an isolated ion with Co contents below 1%, while Co^{2+}-pair complexes gradually increased in the host with a Co content over 2%.

Excitonic luminescence in $Zn_{1-x}Co_xO$ layers is strongly suppressed (Fig. 3.24(a)). The PL intensity of the $Zn_{0.99}Co_{0.01}O$ layer decreased by an order of one thousand compared to the ZnO layer. Although the origin of the remarkable suppression of excitonic luminescence is unclear at present, we provide a possible explanation. A neighbor of the C.B. in $Zn_{1-x}Co_xO$ can be constituted from complex levels such as excitonic states, internal $3d$ transitions, and photoexcited CT levels. Excited carriers may be transferred to intra $3d$ levels and/or the CT levels by nonradiative processes. Nawrocki *et al.* observed energy transfer from excited carriers to Mn^{2+} ions in $Cd_{1-x}Mn_xS$ due to "Auger recombination" [83]. It is important to note that weak PL emissions of $Zn_{1-x}Co_xO$ layers are associated with nonradiative energy transfer to Co (3d) intra-levels. The inset figure shows the PL spectrum of a $Zn_{1-x}Co_xO$ layer with $x = 0.15\%$. The F_AX emission shifted from 3.377 to 3.398 eV due to the wide band gap. The line shape of D^0X is clearly broadened by Co doping because excited carriers experience spatial potentials that are dependent on local atomic fluctuations. The whole spectra were superimposed with n_{th} ($n = 4$–9) resonant Raman scattering (RRS) [84]. Peak positions of the RRS is independent of temperature, correlating clearly with the Frohlich interaction that causes successive emissions of LO phonons, which can be explained by

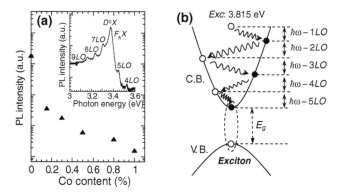

Fig. 3.24. (a) Dependence of integrated PL intensity at the band edge on Co content for $Zn_{1-x}Co_xO$ layers. Inset shows PL spectrum of excitonic transitions at 10 K. (b) Schematic diagram of the cascade mode of OMRRS by a hot exciton-mediated process. The relaxation process is based on the hot exciton involving the quasi-LO phonon

carrier-mediated cascade relaxation (Fig. 3.24(b)). Absorption of the incident photon energy (3.815 eV) is sufficient for the creation of hot electrons. The hot electrons relax to lower energy states with successive emission of LO phonons by cascade events. The RRS at the band edge of $GaA_{1-x}N_x$ (x = 0.001) has been induced by impurity states consisting of excitons bound to isoelectronic nitrogen impurities. Isolated nitrogen centers, which are slightly perturbed by distant nitrogen, result in the RRS and a broadening of exciton bands [85]. This fact suggests that exciton states bound to isolated Co centers in ZnO induced the RRS. Investigations concerning excitonic recombination and nonradiative energy transfer will be enhanced through the use of magneto-PL measurements.

3.4.2 Ferromagnetism

Ferromagnetism (FM) based on Co-doped ZnO spawned a large number of experimental and theoretical studies. The reported experimental results show a broad distribution, indicating that this system is sensitive to preparation methods, measurement techniques, and types of substrate. For example, while some experiments report above room-temperature FM, others report a low FM ordering temperature, or spin-glass or paramagnetic (PM) behavior. Therefore, many researchers have focused extensively on FM properties to elucidate its mechanism. Recent investigations in magneto-optics and magneto-transport in $Zn_{1-x}Co_xO$ layers have reported an interaction between carriers and localized 3d spins [86, 87]. For Sect. 3.4.1, we explained magneto-optics due to the exchange interaction between excitons and localized $3d$ spins. In this session, we introduce a correlation between FM ordering and the concentration of free electrons.

Figure 3.25 shows the logarithmic correlation between magnetization (M_s) at 10 K and electron concentration (n_e) at 300 K for Zn-polar (○) and O-polar (●) $Zn_{0.94}Co_{0.06}O$ layers on ZnO substrates. n_e was controlled by changing $p(O_2)$ from 10^{-4} to 10^{-7} mbar during layer growth. Detailed results of *Ref.* Ms at 10 K correlated closely with n_e at 300 K, indicating that FM ordering is associated with an increase in n_e and represents an essential phenomenon that is independent of the polarity. The inset figure shows M-H hysteresis at 10 K for the $Zn_{0.94}Co_{0.06}O$ layer with a n_e of 1.06×10^{19} cm^{-3}. The saturation magnetization of 112 emu/cm^3 ($1.62\,\mu_B$/Co atom) was obtained in a tetrahedral crystal field. Furthermore, a correlation between Ms and n_e values was found for both polarities. From Fig. 3.25, the relationship shows a scaling behavior of the form $M_s \propto n_e{}^\alpha$, with values of 0.82 and 1.15 for data with Zn- and O-polarities, respectively.

Figure 3.26(a) shows n_e as a function of reciprocal temperature in Zn-polar $Zn_{0.94}Co_{0.06}O$ layers grown under different $p(O_2)$. By reducing $p(O_2)$, n_e at 300 K varied from 10^{16} to 10^{19} cm^{-3}. The carrier activation (E_d) behavior systematically evolves from a linear fitting of n_e near room temperature according to the following formula [88]:

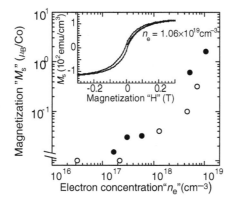

Fig. 3.25. (a) Logarithmic correlation between magnetization (M_s) at 10 K and electron density (n_e) at 300 K for Zn-polar (○) and O-polar (●) $Zn_{0.94}Co_{0.06}O$ layers. *Inset* shows a $M - H$ hysteresis curve for O-polar $Zn_{0.94}Co_{0.06}O$ layers with a n_e of 1.06×10^{19} cm^{-3}

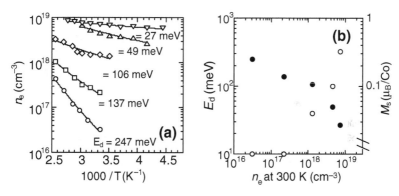

Fig. 3.26. (a) n_e as a function of reciprocal temperature for Zn-polar $Zn_{0.94}Co_{0.06}O$ layers grown under a $p(O_2)$ of 1.4×10^{-4} (○), 1.4×10^{-5} (□), 1.4×10^{-6} (▽), 1.4×10^{-7} (◇), and 6.0×10^{-8} mbar (△). The *solid line* represents the theoretical fit according to (3.7). (b) E_d and M_s at 10 K as a function of n_e at 300 K. (●) and (○) indicate E_d and M_s values, respectively

$$n_e + N_A = \frac{N_D}{1 + g_d \left(\frac{n_e}{N_C}\right) \exp\left(\frac{E_d}{k_B T}\right)} \tag{3.7}$$

where $g_d = g_1/g_0$ represents the factor with degeneracy of the unoccupied donor state $g_0 = 1$ and degeneracy of the occupied donor state $g_1 = 2$, assuming an s-like two-level system, and E_d and N_D represent the donor activation and donor concentration, respectively. For clarity, the quantity N_A represents the total number of deep acceptors acting as electron trapping centers in the host. Furthermore, $N_C = 2(2\pi m^* k_B T/h^2)^{3/2}$ for ZnO denotes the effective density of states in the conduction band, where k_B and h are the Boltzmann and Plank constants, respectively, and m^* represents the electron effective mass in ZnO. E_d decreased from 247 to 49 meV with decreasing $p(O_2)$, and N_D and N_A values were in the order of 10^{19} to 10^{20} cm^{-3} and 10^{17} cm^{-3} in all layers, respectively. The number of free electrons is only controlled by the activation energy of donor levels. Since Co ions in ZnO layers are in the divalent state, doped Co ions are considered a neutral dopant. Thus, n_e cannot be affected by doping with Co ions. One theoretical investigation proposed the formation of Co$_{Zn}$-native defects in ZnO under an oxygen-rich atmosphere [89]. The decrease in n_e with increasing $p(O_2)$ was due to an increase in E_d following the generation of deep donor levels. Figure 3.26(b) shows E_d and M_s at 10 K as a function of n_e at 300 K. A large M_s was observed in the high n_e regions supplied from shallow donor levels, while an increase in E_d was suppressed following FM ordering with a decrease in n_e. This suggests that the high n_e around the Mott transition supplied from shallow donor levels plays an important role in maintaining FM ordering.

Origin of FM ordering cannot be explained by super-exchange interactions with short-range order because FM ordering appears at a low concentration that is 6% below the percolation threshold (17%) associated with nearest-neighbor cation coupling in ZnO [90]. From theoretical aspects, Zn-CoO possesses large exchange splitting. Up-spin states of the $3d$ orbital are fully occupied and were set on the top of the valence band. On the other hand, partially occupied down-spin states are located near the Fermi level. Until now, common models of FM ordering in ZnO DMS materials suggested a strong coupling between magnetic ions and charge carriers in the vicinity of the Fermi level. If donor impurity such as Zni is introduced in the host lattice, shallow donor levels with low E_d are formed directly under the conduction band (C.B.), and is followed by the formation of an impurity band that causes a delocalization of carriers around the Mott transition. The Fermi

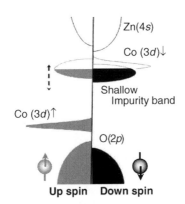

Fig. 3.27. The Fermi level lies in a spin-split impurity band. Schematic format adapted from [91]

level rises near the impurity band, resulting in strong hybridization and charge transfer from the impurity band to the empty $3d$ orbital of Co^{2+} ions near the Fermi level for FM ordering, as shown in Fig. 3.27. This physical origin is derived from the spin-split impurity band model [91] and is suitable for our data in explaining that the transition of magnetic ordering from PM to FM depends on ne. In contrast, deep donor levels in the band gap are strongly localized charge carriers and suppress an itineration of electrons, resulting in low n_e and not FM ordering. Therefore, a hopping motion of free electrons in the impurity band is needed to mediate FM ordering in $Zn_{0.94}Co_{0.06}O$ layers.

3.4.3 Space Separation of Exciton and Localized Spin Systems

$Zn_{1-x}Co_xO$ is of particular interest in studies concerning quantum heterostructures derived from coupling of spins and carriers. However, growth control of a 2D mode has seldom been reported for $Zn_{1-x}Co_xO$ layers despite its practical use in applications such as spintronics. 2D growth for a sharp heterointerface is suitable for layer growth along the Zn-polarity. In session 2, we introduced homoepitaxial growth in ZnO and fabrications of $Mg_xZn_{1-x}O/ZnO$ MQWs based on Zn-polarity. A detailed investigation of the growth mechanism should be performed for Zn-polar $Zn_{1-x}Co_xO$ layers. Quantum wells (QWs) geometry based on $Zn_{1-x}Co_xO/ZnO$ represents spatially separated excitons and localized $3d$ spins at the nano scale. The quantum confinement systems of DMS enhance the exchange interaction and produce an interesting issue from both physics and applications point of view. In our case, electrons and holes were confined in the non-magnetic ZnO layers because the $Zn_{1-x}Co_xO$ layer becomes a barrier layer. This QWs geometry is subjected to the effect of the magnetic spins in the barriers only through wave function penetration. Therefore, this QWs geometry becomes an effective investigation technique of spin-dependent luminescence showing a circular polarization reflecting the FM ordering in $Zn_{1-x}Co_xO$.

The FM ordering of $Zn_{1-x}Co_xO$ layers is due to the Spin-split impurity band mechanism. Therefore, Zn-polar $Zn_{1-x}Co_xO$ layers should be grown in the lower $p(O_2)$ regions below 10^{-6} mbar. However, layer surfaces obtained with this strategy usually possess rough surfaces with many pit holes that are ascribed to threading dislocations in terms of V-pit holes [92]. The strain effect is the primary cause of the formation of the pits, and stacking faults generated by strain relaxation leads to the formation of pits. Thus, these pits cannot be formed to contain the strain energy in the layer below a critical thickness, at which point it is favorable to release the stress by forming pits. Figure 3.28 (a) shows the dependence of pit density on layer thickness for Zn-polar $Zn_{0.94}Co_{0.06}O$ layers grown under a $p(O_2)$ of 10^{-6} and 10^{-7} mbar. Pit density decreased with a decrease in layer thickness and reduced to orders of 10^7 cm^{-2} at a layer thickness of 10 nm. The 2D growth in the $Zn_{0.94}Co_{0.06}O$ layer at the lower $p(O_2)$ regions was obtained whilst maintaining the high n_e required for FM ordering with a decrease in the layer thickness. These

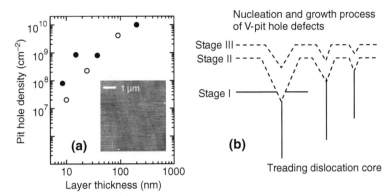

Fig. 3.28. (a) Dependence of pit hole density on layer thickness for Zn-polar $Zn_{0.94}Co_{0.06}O$ layers grown under a $p(O_2)$ of 10^{-6} (•) and 10^{-7} (○) mbar. *Inset* shows an AFM image of the layer with a thickness of 10 nm grown under a $p(O_2)$ of 10^{-6} mbar. (b) Schematic representation of the growth process concerning pit hole defects

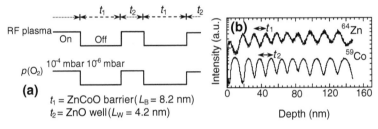

Fig. 3.29. (a) RF plasma power and $p(O_2)$ sequence used for the fabrication of a 10-period $Zn_{0.94}Co_{0.06}O/ZnO$ quantum wells geometry. (b) In-depth profiles of ^{64}Zn and ^{59}Co atoms using SIMS analysis

observations might suggest that a 2D mode in Zn-polarity is retained even at lower $p(O_2)$ within the initial stage of layer growth. An incomplete covering of O atoms on the growing surface due to an oxygen-poor atmosphere results in growth instability of the 2D mode and leads to pit formation. The density of pits rapidly increases as the layer thickness increases due to escalating strain relaxation.

$Zn_{0.94}Co_{0.06}O$ layers possess a high band offset of 156 meV compared with the one of ZnO. This band offset can sufficiently confine electrons and holes in the ZnO wells. We developed periodic oxygen pressure modulated epitaxy to fabricate a QWs geometry. A superlattice (SL) sample with 10 periods of ZnO (4.4 nm)/$Zn_{0.94}Co_{0.06}O$ (8.2 nm) was grown at 400°C under a $p(O_2)$ alternating between 1.4×10^{-4} and 1.4×10^{-6} mbar, corresponding to the ZnO and $Zn_{0.94}Co_{0.06}O$ layers. The RF plasma source was switched on and off during ZnO and $Zn_{0.94}Co_{0.06}O$ growth, respectively Fig. 3.29(a)

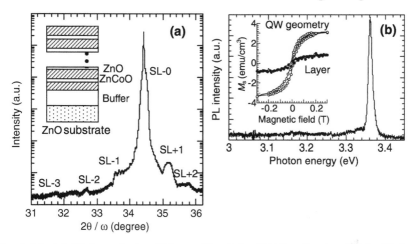

Fig. 3.30. (a) High-resolution $2\theta/\omega$ profile of the (0002) reflection for the SL layer with Zn-polarity. *Inset* shows a schematic cross-section structure of the sample. (b) PL spectrum at 10 K of the SL layer. *Inset* shows $M - H$ hysteresis loops at 10 K for SL klyer (○) and $Zn_{0.06}Co_{0.94}$ layers (●) grown under a $p(O_2)$ of 1.4×10^{-6} mbar

and (b)). SL period was evaluated using Second ion mass spectroscopy (SIMS) analysis (Fig. 3.29(c)). Figure 3.30(a) shows the (0002) XRD pattern of the SL layer. The pronounced fringes and high-order satellite peaks suggest a high crystalline quality as a result of a decrease of imperfection or composition inhomogeneity. The ω-rocking curves of the satellite peaks had very narrow line-widths. Figure 3.30(b) shows the PL spectrum at the band edge of the SL layer. Excitonic emission originating from the ZnO wells was observed at 10 K. Furthermore, a clear hysteresis curve due to FM ordering of the sample was simultaneously obtained at 10 K from the M-H curve (inset of Fig. 3.30(b)). Ferromagnetism and excitonic luminescence were simultaneously obtained by the repetition of the magnetic and non-magnetic layers based on a quantum wells geometry. These results show quantum structures with an interesting coupling between spins and excitons.

3.5 Conclusion

Homoepitaxial growth and MQWs in ZnO along polar and nonpolar directions have been summarized on this chapter. The Zn-polar growth showed atomically flat surfaces, which led to the fabrication of high-quality MQWs with an efficient carrier confinement even at RT. Furthermore, it was found that the spatially separated 3D nanodots were naturally formed on the 2D wetting layers due to the elastic distortion induced by lattice misfit at the $Mg_{0.37}Zn_{0.63}O$/ZnO heterointerface. This S-K forms the basic technology of quantum dots based on ZnO.

On the other hand, anisotropic surface nanowires were self-organized on M-nonpolar ZnO surfaces during homoepitaxial growth. The step-edge barrier effect was related closely to its growth mechanism. MQWs constructed on the nanowires structure allowed to observe highly anisotropic conductivity dependent on the surface morphology, which was similar to quantum wires. This origin was explained as follows. Quantization energy of confined electrons fluctuated due to an interface roughness between the surface nanowires, which strongly restricted electron motion perpendicular to the nanowire arrays.

In Sect. 3.4, we discussed various properties of $Zn_{1-x}Co_xO$. An incorporation of Co atoms enlarged the band gap, and caused huge magneto-optical response. The ferromagnetism in $Zn_{1-x}Co_xO$ layers only appeared when shallow donor levels were formed in the band gap and was explained by the spin-split impurity band model. Finally, we fabricated the $Zn_{1-x}Co_xO/ZnO$ superlattice with a quantum wells geometry using a basic understanding of the growth mechanisms of Zn-polar growth in $Zn_{1-x}Co_xO$. This results in the possibility of spin-dependent photonics based on ZnO.

We believe our demonstrated homoepitaxial technique can be effective for electro-, magneto- and optical applications based on ZnO. We hope that our technique and findings are applied widely with other oxide materials.

References

1. G.A. Hirata, J. Mckittrick, J. Siqueiros, O.A. Lopez, T. Cheeks, O. Contreras, J.Y. Yi, J. Vac. Sci. Technol. A **14**, 791 (1996)
2. S. Masuda, K. Kitamura, Y. Okumura, S. Miyatake, H. Tabata, T. Kawai, J. Appl. Phys. **93**, 1624 (2003)
3. D.G. Thomas, J. Phys. Chem. Solids **15**, 86 (1960)
4. P.Yu, Z.K. Tang, G.K.L. Wong, M. Kawasaki, A. Ohtomo, H. Koinuma, Y. Segawa, Solid State Commun. **103**, 459 (1997)
5. D.M. Bagnall, Y.F. Chen, Z. Zhu, T. Yao, S. Koyama, M.Y. Shen, T. Goto, Appl. Phys. Lett. **70**, 2230 (1997)
6. A.K. Sharma, J. Narayan, J.F. Muth, C.W. Teng, C. Jin, A. Kvit, R.M. Kolbas, O.W. Holland, Appl. Phys. Lett. **75**, 3327 (1999)
7. Y. Chen, H.-J. Ko, S.-K. Hong, T. Sekiuchi, T. Yao, Y. Segawa, J. Vac. Sci. Technol. B **18**, 1514 (2000)
8. T. Makino, C.H. Chia, N.T. Tuan, H.D. Sun, Y. Segawa, M. Kawasaki, A. Ohtomo, K. Tamura, H. Koinuma, Appl. Phys. Lett. **77**, 4250 (2000)
9. G. Coli, K.K. Bajaj, Appl. Phys. Lett. **78**, 2861 (2001)
10. H.D. Sun, T. Makino, Y. Segawa, M. Kawasaki, A. Ohtomo, K. Tamura, H. Koinuma, J. Appl. Phys. **91**, 1993 (2002)
11. A. Tsukazaki, A. Ohtomo, T. Kita, Y. Ohno, H. Ohno, M. Kawasaki, Science **315**, 1388 (2007)
12. J. Stangl, V.Holo, G. Bauer, Rev. Mod. Phys. **76**, 725 (2004)
13. H.P. Schonherr, J. Fricke, Z. Niu, K.J. Friedland, R. Notzel, K.H. Ploog, Appl. Phys. Lett. **72**, 566 (2001)

3 Electro-Magneto-Optics   109

14. F.W. Yan, W.J. Zhang, R.G. Zhang, L.Q. Cui, C.G. Liang, S.Y. Liu, J. Appl. Phys. **90**, 1403 (2001)
15. V.A. Shchukin, D. Bimberg, Rev. Mod. Phys. **71**, 1125 (1999)
16. X.L. Wang, V. Voliotis, J. Appl. Phys. **99**, 121301 (2006)
17. R.D. Vispute, V. Talyansky, Z. Trajanovic. S. Choopum, M. Downes, R.P. Sharma, T. Venkatesan, M.C. Wood, R.T. Lareau, K.A. Jones, A.A. Ilidais, Appl. Phys. Lett. **70**, 2735 (1997)
18. P. Fons, K. Iwata, S. Niki, A. Yamada, K. Matsubara, M. Watanabe, J. Cryst. Growth **209**, 532 (2000)
19. F. Vigue, P. Vennegues, S. Vezian, M. Laugt, J.-P. Faurie, Appl. Phys. Lett. **79**, 194 (2001)
20. T.M. Parker, N.G. Condon, R. Lindsay, F.M. Leibsle, G. Thornton, Surf. Sci. **415**, L1046 (1998)
21. H. Kato, M. Sano, K. Miyamoto, T. Yao, Jpn. J. Appl. Phys. **42**, 2241 (2003)
22. H. Matsui, H. Saeki, T. Kawai, A. Sasaki, M. Yoshimoto, M. Tsubaki, H. Tabata, J. Vac. Sci. Technol. B **22**, 2454 (2004)
23. Z.X. Mei, X.L. Du, Y. Wang, M.J. Ying, Z.Q. Zeng, H. Zheng, J.F. Jia, Q.K. Xue, Z. Zhang, Appl. Phys. Lett. **86**, 112111 (2005).
24. H. Matsui, H. Tabata, Appl. Phys. Lett. **87**, 143109 (2005); J. Appl. Phys. 99, 124307 (2006)
25. H. Ohono, Science **281**, 951 (1998)
26. A.E. Turner, R.L. Gunshor, S. Datta, Appl. Opt. **22**, 3152 (1983)
27. K. Ando, H. Saito, Z. Jin, T. Fukumura, M. Kawasaki, Y. Matsumoto, H. Koinuma, J. Appl. Phys. **89**, 7284 (2001)
28. K. Ueda, H. Tabata, T. Kawai, Appl. Phys. Lett. **79**, 988 (2001)
29. Z.Y. Xiao, H. Matsui, N. Hasuike, H. Harima, H. Tabata, J. Appl. Phys. **103**, 043504 (2008)
30. J.A. Gaj, W. Grieshaber, C.Bodin-Deshayes, J. Cibert, G. Feuillet, Y. Merle d'Aubigné, A. Wasiela, Phys. Rev. B **46**, 5266 (1992)
31. H. Matsui, H. Tabata, Phys. Rev. B **75**, 014438 (2007); Phys. Stat. Sol. (c) 3, 4106 (2006)
32. A.M. Mariano, R.E. Hanneman, J. Appl. Phys. **34**, 384 (1963)
33. O. Dulub, L.A. Boatner, U. Diebold, Surf. Sci. **519**, 201 (2002)
34. M. Katayama, E. Nomura, N. Kanekama, H. Soejima, M. Aono, Nucl. Instrum. Methods Phys. Res. B **33**, 857 (1988)
35. T. Ohnishi, A. Ohtomo, I. Ohkubota, M. Kawasaki, M. Yoshimoto, H. Koinuma, Mater. Sci. Eng. B **56**, 256 (1998)
36. H. Kato, K. Miyamoto, M. Sano, T. Yao, Appl. Phys. Lett. **84**, 4562 (2004)
37. J.S. Park, S.K. Hong, T. Minegishi, S.H. Park, I.H. Im, T. Hanada, M.W. Cho, T. Yao, J.W. Lee, J.Y. Lee, Appl. Phys. Lett. **90**, 201907 (2007)
38. D.C. Reynolds, D.C. Look, B. Jogai, C.W. Litton, T.C. Collins, W. Harsch, G. Cantwell, Phys. Rev. B **57**, 12151 (1998)
39. S.F. Chichibu, T. Onuma, M. Kubota, A. Uedono, T. Sota, A. Tsukazaki, A. Ohtomo, M. Kawasaki, J. Appl. Phys. **99**, 093505 (2006)
40. S.-K. Hong, H.-J. Ko,Y. Chen, T. Yao, J. Vac. Sci. Technol. B **20**, 1656 (2002)
41. H. Maki, I. Sakaguchi, N. Ohashi, S. Sekiguchi, H. Haneda, J. Tanaka, N. Ichinose, Jpn. J. Appl. Phys. **75** (2003)
42. N. Hasuike, H. Fukumura, H. Harima, K. Kisoda, H. Matsui, H. Saeki, H. Tabata, J. Phys. Condens. Matter **16**, S5807 (2004).

43. H. Matsui, H. Tabata, N. Hasuike, H. Harima, J. Appl. Phys. **99**, 024902 (2006)
44. E. Bauer, Z. Kristallogr. **110**, 372 (1958)
45. D.J. Eaglesham, M. Cerullo, Phys. Rev. Lett, **64**, 1943 (1990)
46. T. Walter, C.J. Humphreys, A.G. Cullis, Appl. Phys. Lett. **71**, 809 (1997)
47. H. Matsui, N. Hasuike, H. Harima, T. Tanaka, H. Tabata, Appl. Phys. Lett. **89**, 091909 (2006)
48. J.W. Matthews, S. Mader, T.B. Light, J. Appl. Phys. **41**, 3800 (1970)
49. S.S. Srinivasan, L. Geng, R. Liu, F.A. Ponce, Y. Narukawa, S. Tanaka, Appl. Phys. Lett. **25**, 5187 (2003)
50. A. Ohotomo, M. Kawasaki, I. Ohkubo, H. Koinuma, T. Yasuda, Y. Segawa, Appl. Phys. Lett. **75**, 980 (1999)
51. P.F. Fewster, *X-ray Scattering from Semiconductors* (Imperial College Press, London, 2000)
52. B.P. Zhang, N.T. Binh, K. Wakatsuki, C.Y. Liu, Y. Segawa, N. Usami, Appl. Phys. Lett. **86**, 032105 (2005)
53. N.V. Lomasnov, V.V. Travnikov, S.O. Kognovitskii, S.A. Gurevich, S.I. Nesterov, V.I. Skopina, M. Rabe, F.Heneberger, Phys. Solid State **40**, 1413 (1998)
54. Beltran, J. Andres. M. Calatayud, J.B.L. Martins, Chem. Phys. Lett. **338**, 224 (2001)
55. H. Matsui, N. Hasuike, H. Harima, H. Tabata, J. Appl. Phys. **104**, 094309 (2008)
56. S. Guha, A. Madhukar, K.C. Rajkumar, Appl. Phys. Lett. **57**, 2110 (1990)
57. B. Rottger, M. Hanbucken, H. Neddermeyer, Appl. Surf. Sci. **162/163**, 595 (2000)
58. M.D. Johnson, C. Orme, A.W. Hunt, D. Graff, J. Sudijiono, L.M. Sander, B.G. Orr, Phys. Rev. Lett. **72**, 116 (1994)
59. O. Dulub, L.A. Boatner, U. Diebold, Surf. Sci. **504**, 271 (2002)
60. R.L. Schwoebel, J. Appl. Phys. **37**, 3682 (1966)
61. T. Michely, J. Krug, *Islands, Mounds, Atoms*, Springer Series in Surface Science Vol.42 (Springer, New York, 2004)
62. Y.M. Yu, B.G. Liu, Phys. Rev. B **77**, 195327 (2008)
63. T. Moriyama, S. Fujita, Jpn. J. Appl. Phys. **44**, 7919 (2005)
64. J. Zuniga-Perez, V. Munoz-Sanjose, E. Palacios-Lidon, J. Colchero, Appl. Phys. Lett. **88**, 261912 (2006).
65. Y. C. Park, C. W. Litton. T.C. Collins, D. C. Reynolds, Phys. Rev. B **143**, 512 (1966).
66. D.C. Reynolds, D.C. Look, B. Jagai, C.W. Litton, T.C. Collins, T. Harris, M.J. Callahan, J.S. Bailey, J.Appl. Phys. **86**, 5598 (1999).
67. W.J. Fan, J.B. Xia, P.A. Agus, S.T. Tan, S.F. Yu, X.W. Sun, J. Appl. Phys. **99**, 013702 (2006)
68. D. C. Reynolds, D. C. Look, B. Jogai, C.W. Litton, G. Cantwell, W.C. Harsch, Phys. Rev. B **60**, 2340 (1999)
69. H. Matsui, H. Tabata, Appl. Phys. Lett. **94**, 161907 (2009)
70. A. Mang, K. Reimann, St. Rubenacke, Solid State Commun. **94**, 251 (1999)
71. A. Niwa, T. Ohtoshi, T. Kuroda, Jpn. J. Appl. Phys. **35**, L599 (1996)
72. A. Waag, S. Schmeusser, R.N. Bicknell-Tassius, D.R. Yakovlev, W. Ossau, G. Landwehr, I.N. Uraltsev, Appl. Phys. Lett. **59**, 2995 (1991)
73. R. Notzel, P.H. Ploog, J. Vac. Sci. Technol. A **10**, 617 (1992)
74. H. Sakai, T. Noda, K. Hirakawa, M. Tanaka, T. Matsusue, Appl. Phys. Lett. **51**, 1934 (1987)

75. F .Szmulowicz, S. Elhamri, H.J. Haugan, G.J. Brown, W.C. Mitchel, J. Appl. Phys. **101**, 043706 (2007)
76. K. Sato, H. Katayama-Yoshida, Jpn. J. Appl. Phys. **40**, L334 (2001)
77. M.H. F. Sluiter, Y. Kawazoe, P. Sharma, A. Inoue, A.R. Raju, C. Rout, U.V. Waghmare, Phys. Rev. Lett. **94**, 187204 (2005)
78. H. Zheng, Physica B **212**, 125 (1995)
79. W.K. Liu, G.M. Salley, D.R. Gamelin, J. Phys. Chem. B **109**, 14486 (2005)
80. J. Dreyhsig, B.Litzenburger, Phys. Rev. B **54**, 10516 (1996)
81. R.J. Powell, W.E. Spicer, Phys. Rev. **2**, 2182 (1970)
82. J.K. Furdyna, J. Appl. Phys. **64**, R29 (1988)
83. M. Naweocki, Y.G. Rubo, J.P. Lascaray, D. Coquillat, Phys. Rev. B **52**, R2241 (1995)
84. W.H. Sun, S.J. Chua, L.S. Wang, X.H. Zhang, J. Appl. Phys. **91**, 4917 (2002)
85. H.M. Cheong, Y. Zhang, A. Mascarenhas, J.F. Geisz, Phys. Rev. B **61**, 13687 (2000)
86. K.R. Kittlstved, D.A. Schwartz, A.C. Tuan, S.M. Heald, S.A. Chambers, D.R. Gamelin, Phys. Rev. Lett. **97**, 037203 (2006)
87. J.R. Neal, A.J. Behan, R.M. Ibranhim, H.J. Blythe, M. Ziese, A.M. Fox, G.A. Gehring, Phys. Rev. Lett. **96**, 197208 (2006)
88. R. Schaub, G. Pensl, M. Schulz, C. Holm, Appl. Phys. A **34**, 215 (1984)
89. F. Oba, T. Yamamoto, Y. Ikuhara, I. Tanaka, H. Adachi, Mater. Trans. **43**, 1439 (2002)
90. E.C. Lee, K.J. Chang, Phys. Rev. B **69**, 085205 (2004)
91. J. M. Coey, M. Venkatesan, C.B. Fitzgerald, Nat. Mater. **4**, 173 (2005)
92. T.L. Song, J. Appl. Phys. **98**, 084906 (2005)

4

Nonadiabatic Near-Field Optical Polishing and Energy Transfers in Spherical Quantum Dots

W. Nomura, T. Yatsui, and M. Ohtsu

4.1 Introduction

In the first half of this chapter, a novel fabrication method called nanophotonic polishing is reviewed. This method is a probeless and maskless optical processing technique that employs a nonadiabatic photochemical reaction. Nanophotonics has already brought about innovation in fabrication methods, such as with photochemical vapor deposition [1] and photolithography [2]. Conventional photochemical vapor deposition is a way to deposit materials on a substrate using a photochemical reaction with ultraviolet light that predissociates metal-organic molecules by irradiating gaseous molecules or molecules adsorbed on the substrate. Consequently, the electrons in the molecules are excited to a higher energy level, following the Franck–Condon principle. This is an adiabatic process, which indicates that the Born–Oppenheimer approximation is valid. However, it has been discovered that an optical near field with much lower photon energy (i.e., visible light) can dissociate the molecule. This phenomenon has been explained using a theoretical model of the virtual exciton-polariton exchange between a metal-organic molecule and the fiber probe tip used to generate the optical near field. In other words, this exchange excites not only the electron, but also molecular vibrations. This is a nonadiabatic process, which does not follow the Franck–Condon principle, and so the Born–Oppenheimer approximation is no longer valid.

In the last half of this chapter, optical near-field interactions and energy transfer between spherical quantum dots (QDs) are reviewed. The optical near-field interaction is a short-range interaction mediated by optical electromagnetic fields that enable excitation energy transfer [3–5], which is advantageous in applications for biomolecular imaging, sensing, photonic devices, and nanofabrication. Many approaches have been used to explain the excitation energy transfer via the near-field interaction between nanoscale particles, including QDs, molecules, and metallic nanoparticles. Fluorescence resonance energy transfer based on the Förster mechanism is one of the popular treatments [6]. In this approach, the particle is approximated as an ideal single point dipole or multipole; however, this is insufficient for examining the energy transfer between nano-sized particles because the distance between them is too short. Coherent excitation energy transfer in metallic nanoparticle chains has been analyzed using a polarization density distribution instead of a single electric dipole model. As long as the particles are large and the carriers are not quantized, the polarization density distribution is easily obtained. In the case of QDs, however, the polarization density distribution must be defined from the wave functions of photoexcited carriers because this distribution corresponds to the transition dipole distribution in two-level QDs. Therefore, we define the effective dipoles of the optical source and optical absorber for QDs (i.e., a donor and an acceptor). An additional problem is that the interaction between QDs in the host matrix is complicated because the polarization is induced simultaneously in the host matrix by a donor QD, making it difficult to obtain the polarization distribution in the host matrix. In our approach, we describe the optical near-field interaction between QDs as exciton-polariton tunneling to overcome these difficulties [7–10].

4.2 Nanophotonic Polishing Using a Nonadiabatic Photochemical Reaction

An ultra-flat substrate (sub-nanometer scale roughness) is required for the manufacture of high-quality, extreme UV optical components, high-power lasers, and ultrashort-pulse lasers, plus future photonic devices at the sub-100-nm scale. It is estimated that the required surface roughness, R_a, will be less than 1 Å [11]. This R_a value is an arithmetic average of the absolute values of the surface height deviations measured from a best-fit plane, and is given by

$$R_a = \frac{1}{l} \int_0^l |f(x)| dx$$
$$\cong \frac{1}{n} \sum_{i=1}^n |f(x_i)| \qquad (4.1)$$

where $|f(x_i)|$ are absolute values measured from the best-fit plane and l is the evaluation length. Physically, dx corresponds to the spatial resolution of the measurement of $f(x)$, and n is the number of pixels in the measurement ($n = l/dx$). Conventionally, chemical-mechanical polishing (CMP) is used to achieve flat surfaces [12]. However, with CMP, it is difficult to reduce R_a to less than 2 Å, as the polishing pad roughness is typically around 10 μm and the diameters of the polishing particles in the slurry are as large as 100 nm. In addition, polishing causes scratches or digs due to the contact between the polishing particles and/or impurities in the slurry and the substrate.

Our interest in applying an optical near field to nanostructure fabrication was generated because of its high-resolution capability – beyond the diffraction limit – and because of its novel photochemical properties, whereby the reaction is classified as nonadiabatic due to its energy transfer via a virtual exciton-phonon-polariton [1, 13]. In this chemical vapor deposition, photodissociation of the molecules is driven by the light source at a lower photon energy than the molecular absorption edge by a multiple-step excitation via vibrational energy levels [14]. Following this process, we propose a novel method of polishing using nonadiabatic optical near-field etching.

4.2.1 Nonadiabatic Optical Near-Field Etching

A continuous wave laser ($\lambda = 532\,\text{nm}$) was used to dissociate the Cl_2 gas through a nonadiabatic photochemical reaction. The photon energy is smaller than that corresponding to the absorption edge of Cl_2 ($\lambda = 400\,\text{nm}$) [15], so the Cl_2 adiabatic photochemical reaction is avoided. However, because the substrate has nanometer-scale surface roughness, the generation of a strong optical near field on the surface is expected from simple illumination, with no focusing required (Fig. 4.1a). Since a virtual exciton-phonon-polariton can be excited on this roughness, a higher molecular vibrational state can be excited than on the flat part of the surface, where there is no virtual exciton-phonon-polariton. Cl_2 is therefore selectively photodissociated wherever the optical near field is generated. These dissociated Cl_2 molecules then etch away the surface roughness, and the etching process automatically stops when the surface becomes flat (Fig. 4.1b).

4.2.2 Experiment

We used 30-mm-diameter planar synthetic silica substrates built by vapor-phase axial deposition with an OH group concentration of less than 1 ppm [16]. The substrates were preliminarily polished by CMP prior to the nonadiabatic optical near-field etching, which was performed at a Cl_2 pressure of 100 Pa at room temperature with a continuous wave laser ($\lambda = 532\,\text{nm}$) having a uniform power density of $0.28\,\text{W/cm}^2$ (see Fig. 4.2a). Surface roughness was

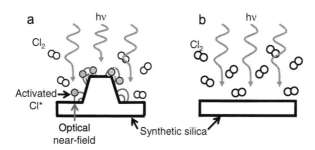

Fig. 4.1. Schematic of the near-field etching (**a**) during the etching process and (**b**) after etching

116   W. Nomura et al.

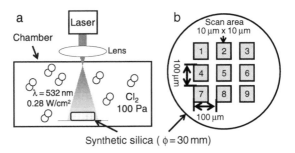

Fig. 4.2. Schematic of (**a**) the experimental setup and (**b**) the AFM measurement

evaluated using an atomic force microscope (AFM). Since the scanning area of the AFM was much smaller than the substrate, we measured the surface roughness, R_a, in nine representative areas, each $10\,\mu m \times 10\,\mu m$, separated by $100\,\mu m$ (see Fig. 4.2b). The scanned area was 256×256 pixels with a spatial resolution of $40\,nm$. The average value, $\bar{R}_a$, of the nine R_a values obtained before the nonadiabatic optical near-field etching and evaluated through the AFM images, was $2.36 \pm 0.02\,\text{Å}$. We cleaned the substrate ultrasonically using deionized water and methanol before and after the nonadiabatic optical near-field etching.

4.2.3 Results and Discussion

Figure 4.3a and b shows typical AFM images of the silica substrate area before and after nonadiabatic optical near-field etching, respectively. Note that the surface roughness was drastically reduced, as supported by the cross-sectional profiles along the dashed white lines in Fig. 4.3a and b (see Fig. 4.3c). We found a dramatic decrease in the value of the peak-to-valley roughness from $1.2\,nm$ (dashed curve) to $0.5\,nm$ (solid curve). Furthermore, note that the scratch seen in the AFM image before nonadiabatic optical near-field etching has disappeared. This indicates that rougher areas of the substrate had a higher etching rate, possibly because of greater intensity of the optical near field, leading to a uniformly flat surface over a wide area.

Figure 4.4a shows the etching time dependence of $\bar{R}_a$. We found that $\bar{R}_a$ decreases as the etching time increases. The minimum $\bar{R}_a$ was $1.37\,\text{Å}$ at an etching time of $120\,min$, while the minimum R_a among the nine areas was $1.17\,\text{Å}$. Because the process is performed in a sealed chamber, the saturation in the decrease of $\bar{R}_a$ might originate from the decrease in the Cl_2 partial pressure during etching. A further decrease in $\bar{R}_a$ would be expected under constant Cl_2 pressure. Figure 4.4b shows the time dependence of the standard deviation of R_a (ΔR_a), which was obtained in one scanning area. We found a dramatic decrease in ΔR_a after $60\,min$, although we also found an increase in ΔR_a in the early stages of nonadiabatic optical near-field etching. This might have been caused by impurities, such as OH, on the substrate surface.

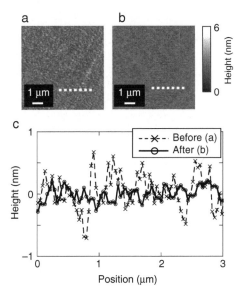

Fig. 4.3. Typical AFM images of the silica substrate (**a**) before and (**b**) after nonadiabatic optical near-field etching. (**c**) Cross-sectional profiles along the *white dotted lines* in (**a**) and (**b**). The *dashed curve* with cross marks and the *solid curve* with *open circles* correspond to the profiles before and after etching, respectively

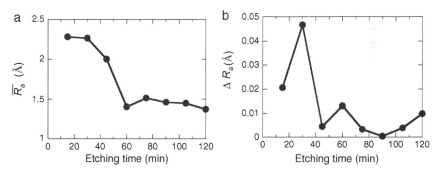

Fig. 4.4. The etching time dependence of (**a**) the average R_a ($\bar{R}_a$) and (**b**) the standard deviation of R_a (ΔR_a)

4.3 Optical Near-Field Energy Transfer Between Spherical Quantum Dot Systems

Systems of optically coupled quantum-size structures should be applicable to quantum information processing [17, 18]. Additional functional devices (i.e., nanophotonic devices [3, 19–21]) can be realized by controlling the excitonic excitation in QDs and quantum wells (QWs). This section reviews the recent achievements with nanophotonic devices based on spherical QDs.

4.3.1 Exciton Energy Levels in Spherical Quantum Dots

The translational motion of the exciton center of mass is quantized in nanoscale semiconductors when the size is as small as an exciton Bohr radius. If the QDs are assumed to be spheres having radius R, with the following potential

$$V(x) = \begin{cases} 0 & \text{for } |x| \le R \\ \infty & \text{for } |x| > R \end{cases} \tag{4.2}$$

then the quantized energy levels are given by a spherical Bessel function as

$$R_{nl}(r) = A_{nl} j_l \left(\rho_{n,l} \frac{r}{R} \right) \tag{4.3}$$

Figure 4.5 shows the lth order of the spherical Bessel function. Note that an odd quantum number of l has an odd function and it is a dipole-forbidden energy state. To satisfy the boundary conditions as

$$R_{nl}(R) = A_{nl} j_l(\rho_{n,l}) \tag{4.4}$$

the quantized energy levels are calculated using

$$E(n, l) = E_B + \frac{\bar{h}^2 \pi^2}{2mR^2} \xi_{n,l}^2 \tag{4.5}$$

where $\pi \xi_{n,l} = \rho_{n,l}$ is the nth root of the spherical Bessel function of the lth order. The principal quantum number, n, and the angular momentum quantum number, l, take values $n = 1, 2, 3 \dots$ and $l = 0, 1, 2 \dots$, while $\xi_{n,l}$ takes values $\xi_{1,0} = 1$, $\xi_{1,1} = 1.43$, and $\xi_{1,2} = 1.83$, $\xi_{2,0} = 2$, and so on (see Table 4.1) [22].

Figure 4.6 shows schematic drawings of different-sized spherical QDs (QDS and QDL) and the confined exciton energy levels. Here, R and $1.43\,R$ are the radii of spherical QDs, QDS and QDL, respectively. According to (4.4), the quantized exciton energy levels of $E(1,0)$ in QDS (E_{S1}) and $E(1,1)$ in QDL (E_{L2}) are in resonance. Although the energy state $E(1,1)$ is a dipole-forbidden state, the optical near-field interaction is finite for such coupling to the forbidden energy state [23]. In addition, the inter-sublevel transition time τ_{sub}, from higher exciton energy levels to the lowest one, is generally less than a few picoseconds and is much shorter than the transition time due to optical near-field coupling (τ_t) [24]. Therefore, most of the energy of the exciton in a QDS with radius R transfers to the lowest exciton energy level in the neighboring QDL with a radius of $1.43\,R$, and recombines radiatively at the lowest level. In this manner, unidirectional energy flow is achieved.

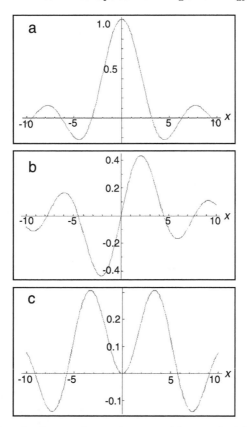

Fig. 4.5. lth order of spherical Bessel function. (a) $l = 0 (j_0(x) = \frac{\sin x}{x})$, (b) $l = 1 (j_1(x) = \frac{1}{x}(\frac{\sin x}{x} - \cos x))$, and (c) $l = 2 (j_2(x) = \frac{1}{x}(\frac{3-x^2}{x^2}) \sin x - \frac{3}{x} \cos x)$

Table 4.1. Calculated $\xi_{n,l}$ to satisfy the condition of $j_1(\pi \xi_{n,l})$

	$l = 0$	$l = 1$	$l = 2$	$l = 3$	$l = 4$	$l = 5$
$n = 1$	3.14	4.49	5.76	6.99	8.18	·
$n = 2$	6.26	7.73	9.10	10.42	·	·
$n = 3$	9.42	·	·	·	·	·
$n = 4$	·	·	·	·	·	·

4.3.2 Resonant Energy Transfer Between CdSe QDs

To evaluate the energy transfer and the energy dissipation, we used CdSe/ZnS core-shell QDs from *Evident Technologies*. As described in the adjacent subsection, assuming that the respective diameters, ϕ, of the QDS and QDL were

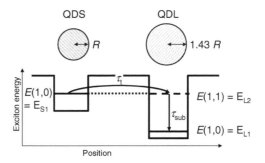

Fig. 4.6. Schematic drawings of different-sized spherical QDs (QDS and QDL) and the confined-exciton energy levels

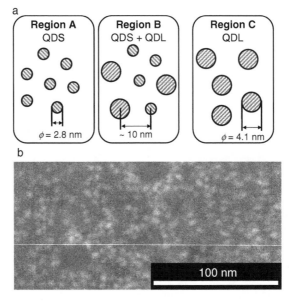

Fig. 4.7. (a) Schematic images of CdSe QDs dispersed substrate. Regions A, B, and C are covered by QDSs, both QDSs and QDLs, and QDLs, respectively. (b) TEM image of dispersed CdSe/ZnS core-shell QDs in region B

2.8 and 4.1 nm, the ground energy level in the QDS and the excited energy level in the QDL resonate [25]. A solution of QDSs ($\phi = 2.8$ nm) and QDLs ($\phi = 4.1$ nm) in 1-feniloctane at a density of 1.0 mg/mL was dropped onto a mica substrate (see Fig. 4.7a), such that regions A and C consisted of QDSs and QDLs, respectively, while there were both QDSs and QDLs in region B. Using transmission electron microscopy (TEM), we confirmed that the mean center-to-center distance of each QD was maintained at less than about 10 nm in all regions due to the 2-nm-thick ZnS shell and surrounding ligands (2-nm-length long chain amine) of the QDs (Fig. 4.7b).

In the following experiments, the light source used was the third harmonic of a mode-locked Ti:sapphire laser (photon energy $h\nu = 4.05\,\text{eV}$, frequency 80 MHz, and pulse duration 2 ps). The incident power of the laser was 0.6 mW and the spot size was $1 \times 10^{-3}\,\text{cm}^2$. The density of QDs was less than $3.5 \times 10^{12}\,\text{cm}^{-2}$, and the quantum yield of CdSe/ZnS QD was 0.5. Under these conditions, the probability of exciton generation by one laser pulse in each QD was calculated to be 1.6×10^{-2}. Therefore, we assumed that single-exciton dynamics apply in the following experiments.

The energy transfer was confirmed using micro-photoluminescence (PL) spectroscopy. Temperature-dependent micro-PL spectra were obtained. In the spectral profile of the PL emitted from region A, we found a single peak that originated from E_{S1} at a photon energy of 2.29 eV, between room temperature and 60 K. From region C, the single peak, which originated from E_{L1}, was found at $h\nu = 2.07\,\text{eV}$. In contrast, region B had two peaks at room temperature, as shown in Fig. 4.8. This figure also shows that the PL intensity peak at $h\nu = 2.29\,\text{eV}$ decreased relative to that at $h\nu = 2.07\,\text{eV}$ on decreasing the temperature. This relative decrease in the PL intensity was due to the energy transfer from E_{S1} to E_{L2} and the subsequent rapid dissipation to E_{L1}. This is because the coupling between the resonant energy levels becomes stronger due to the increase in the exciton decay time on decreasing the temperature [26]. Furthermore, although nanophotonic device operation using CuCl quantum

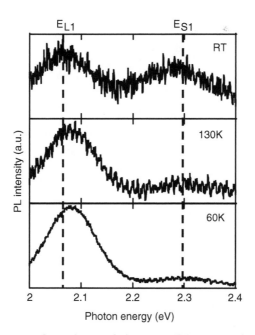

Fig. 4.8. Temperature dependence of the micro-PL spectra from region B. The *dashed line* shows the energy level of E_{S1} and E_{L1}

cubes [19–21] and ZnO QWs [27] has been reported at 15 K, we observed a decrease in the PL intensity at $h\nu = 2.29$ eV at temperatures as high as 130 K, which is advantageous for the higher-temperature operation of nanophotonic devices.

To confirm this energy transfer from QDS to QDL at temperatures under 130 K, we evaluated the dynamic property of the energy transfer using time-resolved spectroscopy and applying a time-correlated single-photon counting method. Circles A_S, squares B_S, and triangles B_L in Fig. 4.9a represent the respective time-resolved micro-PL intensities (at 60 K) from E_{S1} in region A, E_{S1} in region B, and E_{L1} in region B. The peak intensities at $t = 0$ were normalized to unity. Note that B_S decreased faster than A_S, although these signals were generated from QDs of the same size. In addition, although the exciton lifetime decreases on increasing the QD size, owing to the increased oscillator strength, B_L decreased more slowly than A_S over the range $t < 0.2$ ns (see the inset of Fig. 4.9a). Furthermore, as we did not see any peak in the power spectra of A_S, B_S, and B_L, we believe that the temporal signal changes originated from the optical near-field energy transfer and subsequent dissipa-

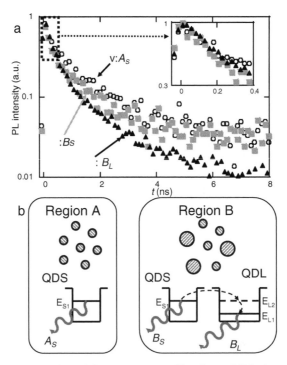

Fig. 4.9. (a) Time-resolved PL intensity profiles from QDSs in region A (*open circles A_S*), QDSs in region B (*gray squares B_S*), and QDLs in region B (*black triangles B_L*). The peak intensities were normalized at $t = 0$. (b) Schematic of the respective system configurations in regions A and B

tion. Since the QDSs in region B were near QDLs whose excited energy level resonates with E_{S1} (see Fig. 4.9b), near-field coupling between the resonant levels resulted in the energy transfer from the QDS to the QDL and the consequent faster decrease in the excitons of the QDS in region B compared to region A. Furthermore, as a result of inflow of the carriers from the QDS to the QDL, the PL intensity from the QDL near the QDS decayed more slowly than that of the QDS.

For comparison, we also obtained time-resolved PL profiles for different pairs of CdSe/ZnS QDs. Their diameters were $\phi = 2.8\,\mathrm{nm}$ (QDS) and $3.2\,\mathrm{nm}$ (QDM), which means that their energy levels were not resonant with each other. Figure 4.10a shows a schematic of a sample named region D, where QDSs and QDMs are mixed with a mean center-to-center distance of less than $10\,\mathrm{nm}$. Circles A_S and cross marks D_S in Fig. 4.10b show the time-resolved PL intensity (at $30\,\mathrm{K}$) due to the ground energy level in QDSs from regions A and D, respectively. No difference was seen in the decay profiles, which indicates that the excited carriers in QDSs did not couple with QDMs due to being off-resonance, and consequently, no energy was transferred (Fig. 4.10c). This supports the idea that Figs. 4.8 and 4.9 demonstrate energy transfer and subsequent dissipation due to near-field coupling between the resonant energy levels of the QDSs and QDLs.

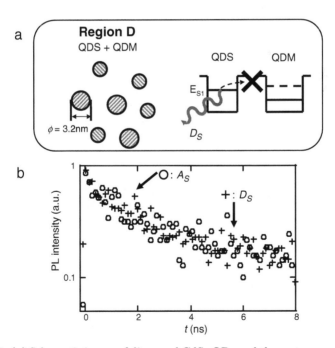

Fig. 4.10. (a) Schematic image of dispersed CdSe QDs and the system configuration in region D. (b) Time-resolved PL intensity profiles from QDSs in region A (*open circles A_S*) and QDSs in region D (*cross marks D_S*). The peak intensities were normalized at $t = 0$

To evaluate the exciton energy transfer from QDS to QDL quantitatively, we investigated the exciton dynamics by fitting multiple exponential decay curve functions to curves A_S, B_S, and B_L [28, 29]:

$$A_S = R_{S1} \exp \frac{-t}{\tau_{S1}} + R_{S2} \exp \frac{-t}{\tau_{S2}}, \qquad (4.6)$$

$$B_S = R_S \cdot A_S + R_t \exp \frac{-t}{\tau_t}, \qquad (4.7)$$

and

$$B_L = R_{L1} \exp \frac{-t}{\tau_{L1}} + R_{L2} \exp \frac{-t}{\tau_{L2}}, \qquad (4.8)$$

We used a double-exponential decay for A_S and B_L (4.6) and (4.8), which corresponds to the non-radiative lifetime (fast decay: τ_{S1} and τ_{L1}) and radiative lifetime of free-carrier recombinations (slow decay: τ_{S2} and τ_{L2}). Given the imperfect homogeneous distribution of the QDSs in region B, some QDSs lacked energy transfer routes to QDLs. However, we introduced the mean energy transfer time, τ_t, from QDSs and QDLs in (4.6). In these equations, we neglected the energy dissipation time τ_{sub} of about 1 ps [30] because that is much smaller than exciton lifetimes and energy transfer times. Figure 4.11 shows the best-fit numerical results and experimental data. Here, we used exciton lifetimes of $\tau_{S2} = 2.10$ ns and $\tau_{L2} = 1.79$ ns. The mean energy transfer time was $\tau_t = 135$ ps, which is comparable to the observed energy transfer

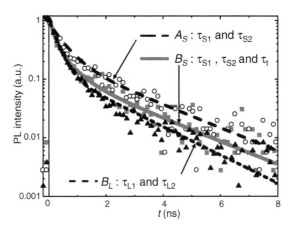

Fig. 4.11. Experimental results (*open circles, gray squares, and black triangles*) and *fitted curves* (*black broken, gray solid, and black short broken curves*) using (4.6), (4.7) and (4.8) for the PL intensity profiles. The fitting parameters were $R_{S1} = 0.560, \tau_{S1} = 2.95 \times 10^{-10}, R_{S2} = 0.329, \tau_{S2} = 2.10 \times 10^{-9}, R_S = 0.740,$ $R_t = 0.330, \tau_t = 1.35 \times 10^{-10}, R_{L1} = 0.785, \tau_{L1} = 2.94 \times 10^{-10}, R_{L2} = 0.201,$ $\tau_{L2} = 1.79 \times 10^{-9}$

time (130 ps) in CuCl quantum cubes [19] and ZnO QW structures [27]. Furthermore, the relation $\tau_t < \tau_{S2}$ agrees with the assumption that most of the excited excitons in QDSs transfer to E_{L2} in a QDLs before being emitted from the QDS.

4.3.3 Control of the Energy Transfer Between ZnO QDs

ZnO is a promising material for room temperature operation of nanophotonic devices because of its large exciton-binding energy [31–33]. Here, we used chemically synthesized ZnO QDs to realize a highly integrated nanophotonic device. We observed the energy transfer from smaller ZnO QDs to larger QDs with mutually resonant energy levels. The energy transfer time and energy transfer ratio between the two QDs were also calculated from the experimental results [34].

ZnO QDs were prepared using the sol-gel method [35, 36] as follows.

- A sample of 1.10 g (5 mmol) of $Zn(Ac)_2 \cdot 2H_2O$ was dissolved in 50 mL of boiling ethanol at atmospheric pressure, and the solution was then immediately cooled to 0 °C. A sample of 0.29 g (7 mmol) of $LiOH \cdot H_2O$ was dissolved in 50 mL of ethanol at room temperature in an ultrasonic bath and cooled to 0 °C. The hydroxide-containing solution was then added dropwise to the $Zn(Ac)_2$ suspension with vigorous stirring at 0 °C. The reaction mixture became transparent after approximately 0.1 g of LiOH had been added. The ZnO sol was stored at 0 °C to prevent particle growth.
- A mixed solution of hexane and heptane, with a volume ratio of 3:2, was used to remove the reaction products (LiAc and H_2O) from the ZnO sol.
- To initiate particle growth, the ZnO solution was warmed to room temperature. The mean diameter of the ZnO QDs was determined from the growth time, T_g.

Figure 4.12a shows a TEM image of synthesized ZnO dots after the second step. The dark areas correspond to the ZnO QDs. This image suggests that monodispersed single crystalline particles were obtained.

To check the optical properties and diameters of our ZnO QD, we measured the PL spectra using He-Cd laser ($h\nu = 3.81$ eV) excitation at 5 K. We compared the PL spectra of ZnO QD with $T_g = 0$ and $T_g = 42$ h (solid and dashed curves in Fig. 4.12b, respectively). A redshifted PL spectrum was obtained, indicating an increase in the diameter of the QDs. Figure 4.12c shows the growth time dependence of the diameter of the QDs. This was determined from the effective mass model, with peak energy levels in the PL spectra, $E_g^{bulk} = 3.35$ eV, $m_e = 0.28$, $m_h = 1.8$, and $\epsilon = 3.7$ [37]. This result indicates that the growth rate at room temperature was 1.1 nm per day.

Assuming that the diameters, ϕ, of the QDS and the QDL were 3.0 and 4.5 nm, respectively, E_{S1} in the QDS and E_{L2} in the QDL resonated (Fig. 4.13a). An ethanol solution of QDS and QDL was dropped onto a sap-

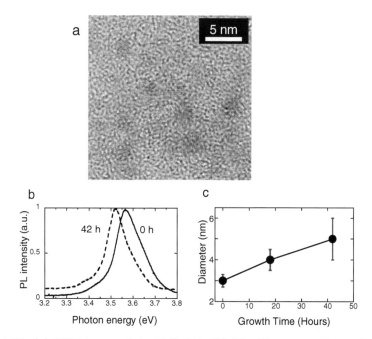

Fig. 4.12. (**a**) TEM image of the ZnO QD. (**b**) The PL spectra observed at 5 K. The *solid* and *dashed curves* indicate growth time $T_g = 0$ and 42 h, respectively. (**c**) The growth time dependence of the mean ZnO QD diameter

phire substrate, and the mean surface-to-surface separation of the QD was found to be approximately 3 nm.

The black solid curve and dashed curve in Fig. 4.13b correspond to the PL spectra of QDS and QDL, with spectral peaks of 3.60 and 3.44 eV, respectively. The gray solid curve in Fig. 4.13b shows the spectrum from the QDS and QDL mixture with $R = 1$, where R is the ratio (number of QDSs)/(number of QDLs). The spectral peak of 3.60 eV, which corresponded to the PL from the QDS, was absent from this curve. This peak was thought to have disappeared due to energy transfer from the QDS to the QDL because the first excited state of the QDL resonated with the ground state of the QDS. Our hypothesis was supported by the observation that when R increased by a factor of eight, the spectral peak from the QDS reappeared (see the black dashed curve in Fig. 4.13b).

To confirm this energy transfer from the QDS to the QDL at 5 K, we evaluated dynamic effects with time-resolved spectroscopy using the time-correlated single-photon counting method. The light source used was the third harmonic of a mode-locked Ti:sapphire laser (photon energy 4.05 eV, frequency 80 MHz, and pulse duration 2 ps). We compared the signals from mixed samples with ratios $R=2$, 1, and 0.5. The curves T_A ($R = 2$), T_B

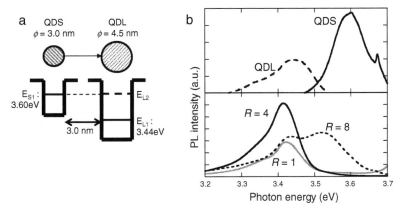

Fig. 4.13. (a) Schematic of the energy diagram between a QDS and QDL. (b) The PL spectra observed at 5 K. *Gray solid curve*, *black solid curve*, and *black dashed curve* indicate mixes with R- ratios of 1, 4, and 8, respectively

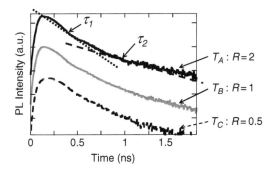

Fig. 4.14. Time-resolved PL spectra observed at 5 K. The values of R were 2, 1, and 0.5 for curves T_A (*black solid curve*), T_B (*gray solid curve*), and T_C (*black dashed curve*), respectively

($R = 1$), and T_C ($R = 0.5$) in Fig. 4.14 show the respective time-resolved PL intensities from the ground state of the QDS (E_{S1}) at 3.60 eV. We investigated the exciton dynamics quantitatively by fitting multiple exponential decay functions [28, 29]:

$$TRPL = A_1 \exp \frac{-t}{\tau_1} + A_2 \exp \frac{-t}{\tau_2} \tag{4.9}$$

We obtained average τ_1- and τ_2-values of 144 and 443 ps, respectively (see Table 4.2). Given the disappearance of the spectral peak at 3.60 eV in the PL spectra, these values likely correspond to the energy transfer time from the QDS to the QDL and the radiative decay time from the QDS, respectively.

Table 4.2. Dependence of the time constants (τ_1 and τ_2) on R as derived from the two exponential fits of the time-resolved PL signals and the coefficient ratio A_1/A_2

$R =$QDS /QDL	τ_1 [ps]	τ_2 [ps]	A_1/A_2
2	133	490	12.4
1	140	430	13.7
0.5	160	410	14.4
Average	144	443	

This hypothesis was supported by the observation that the average value of τ_1 (144 ps) was comparable to the observed energy transfer time in CuCl quantum cubes (130 ps) [19].

We also investigated the value of the coefficient ratio A_1/A_2 (see Table 4.2); this ratio was inversely proportional to R, and hence proportional to the number of QDLs. This result indicated that an excess QDL caused energy transfer from QDS to QDL, instead of direct emission from the QDS.

4.4 Conclusion

We proposed a new polishing method that uses near-field etching based on a nonadiabatic process, with which we obtained an ultra-flat silica surface with a minimum roughness of 1.37 Å. We believe that our technique is applicable to a variety of substrates, including amorphous and crystal ones. Since this technique is a noncontact method without a polishing pad, it can be applied not only to flat substrates but also to three-dimensional substrates that have convex or concave surfaces, such as micro-lenses and the inner-wall surface of cylinders. Furthermore, this method is compatible with mass production.

We observed the dynamic properties of excitonic energy transfer and dissipation between CdSe/ZnS core-shell QDs and ZnO QDs via an optical near-field interaction using time-resolved PL spectroscopy. We observed the dynamic properties of excitonic energy transfer and dissipation between CdSe/ZnS core-shell QDs and ZnO QDs via an optical near-field interaction using time-resolved PL spectroscopy. We experimentally confirmed that optical near-field coupling does not occur between nonresonant energy levels. Furthermore, we successfully increased the energy transfer ratio between the resonant energy states, instead of the radiative decay from the QD. Chemically synthesized spherical nanocrystals, both semiconductor QDs and metallic nanocrystals [38], are promising nanophotonic device candidates because they have uniform sizes, controlled shapes, defined chemical compositions, and tunable surface chemical functionalities.

Acknowledgment

The work in Sect. 4.2 is supported by New Energy and Industrial Technology Development Organization (NEDO) Special Courses: A comprehensive activity for personnel training and industry-academia collaboration based on NEDO projects.

The works in Sect. 4.3 are supported in partial by the Global Center of Excellence (G-COE) "Secure-Life Electronics" sponsored by the Ministry of Education, Culture, Sports, Science and Technology (MEXT), Japan.

References

1. T. Kawazoe, K. Kobayashi, S. Takubo, M. Ohtsu: J. Chem. Phys. **122**, 024715 (2005)
2. T. Kawazoe, K. Kobayashi, S. Sangu, M. Ohtsu, A. Neogi, Unique properties of optical near field and their applications to nanophotonics, in: *Progress in Nano-Electro-Optics V*, ed. by M. Ohtsu (Springer-Verlag, Berlin, 2006)
3. M. Ohtsu, K. Kobayashi, T. Kawazoe, S. Sangu, T. Yatsui, IEEE J. Sel. Top. Quant. Electron. **14**, 1404 (2008)
4. K. Kobayashi, S. Sangu, H. Ito, M. Ohtsu, Phys. Rev. A, **63**, 013806 (2001)
5. T. Kawazoe, K. Kobayashi, J. Lim, Y. Narita, M. Ohtsu, Phys. Rev. Lett. **88**, 067404 (2002)
6. B. W. van der Meer, G. Coker III, S. Y. S. Chen, *Resonant Energy Transfer: Theory and Data* (VCH, New York, 1994)
7. K. Kobayashi, S. Sangu, T. Kawazoe, M. Ohtsu, J. Lumin. **112**, 117 (2005)
8. D. P. Craig, T. Thirunamachandran, *Molecular Quantum Electrodynamics* (Academic, London, 1984)
9. E. Hanamura, Phys. Rev. B, **37**, 1273 (1988)
10. S. Sangu, K. Kobayashi, A. Shojiguchi, M. Ohtsu, Phys. Rev. B, **69**, 115334 (2004)
11. B. Wua, A. Kumar, J. Vac. Sci. Technol. B **25**, 1743 (2007)
12. L. M. Cook, J. Non-Cryst. Solids **120**, 152 (1990)
13. K. Kobayashi, T. Kawazoe, M. Ohtsu, IEEE Trans. Nanotechnol. **4**, 517 (2005)
14. T. Kawazoe, Y. Yamamoto, M. Ohtsu, Appl. Phys. Lett. **79**, 1184 (2001)
15. R. Kullmer, D. Bäuerle, Appl. Phys. A **43**, 227 (1987)
16. T. Izawa, N. Inagaki, Proc. IEEE **68**, 1184 (1980)
17. M. Bayer, P. Hawrylak, K. Hinzer, S. Fafard, M. Korkusinski, Z. R. Wasilewski, O. Stern, A. Forchel, Science **291**, 451 (2001)
18. E. A. Stinaff, M. Scheibner, A. S. Bracker, I. V. Ponomarev, V. L. Korenev, M. E. Ware, M. F. Doty, T. L. Reinecke, D. Gammon, Science **311**, 636 (2006)
19. T. Kawazoe, K. Kobayashi, S. Sangu, M. Ohtsu, Appl. Phys. Lett. 82, **2957** (2003)
20. T. Kawazoe, K. Kobayashi, M. Ohtsu, Appl. Phys. Lett. **86**, 103102 (2005)
21. T. Kawazoe, K. Kobayashi, K. Akahane, M. Naruse, N. Yamamoto, M. Ohtsu, Appl. Phys. B **84**, 243 (2006)
22. N. Sakakura, Y. Masumoto, Phys. Rev. B **56**, 4051 (1997)
23. K. Kobayashi, S. Sangu, H. Itoh, M. Ohtsu, Phys. Rev. A **63**, 013806 (2000)

24. T. Suzuki, T. Mitsuyu, K. Nishi, H. Ohyama, T. Tomimasu, Appl. Phys. Lett. **69**, 4136 (1996)
25. C. Trallero-Giner, A. Debernardi, M. Cardona, M. Menendez-Proupin, A. I. Ekimov, Phys. Rev. B **57**, 4664 (1998)
26. T. Itoh, M. Furumiya, T. Ikehara, C. Gourdon, Solid State Comm. **73**, 271 (1980)
27. T. Yatsui, S. Sangu, T. Kawazoe, M. Ohtsu, S. J. An, J. Yoo, G.-C. Yi, Appl. Phys. Lett. **90**, 223110 (2007)
28. S. A. Crooker, T. Barrick, J. A. Hollingsworth, V. I. Klimov, Appl. Phys. Lett. **82**, 2793 (2003)
29. M. G. Bawendi, P. J. Carroll, L. W. William, L. E. Brus, J. Chem. Phys. **96**, 946 (1992)
30. P. Guyot-Sionnest, M. Shim, C. Matranga, M. Hines, Phys. Rev. B **60**, R2181 (1999)
31. A. Ohtomo, K. Tamura, M. Kawasaki, T. Makino, Y. Segawa, Z. K. Tang, G. K. L. Wong, Appl. Phys. Lett. **77**, 2204 (2000)
32. M. H. Huang, S. Mao, H. Feick, Science **292**, 1897 (2001)
33. H. D. Sun, T. Makino, Y. Segawa, M. Kawasaki, A. Ohtomo, K. Tamura, H. Koinuma, J. Appl. Phys. **91**, 1993 (2002)
34. T. Yatsui, H. Jeong, M. Ohtsu, Appl. Phys. B **93**, 199 (2008)
35. E. A. Meulenkamp, J. Phys. Chem. B **102**, 5566 (1998)
36. L. Spanhel, M. A. Anderseon, J. Am. Chem. Soc. **113**, 2826 (1991)
37. L. E. Brus, J. Chem. Phys. **80**, 4403 (1984)
38. M. Brust, C. J. Kiely, Colloids Surf. A **202**, 175 (2002)

5

Shape-Engineered Nanostructures for Polarization Control in Optical Near- and Far-Fields

M. Naruse, T. Yatsui, T. Kawazoe, H. Hori, N. Tate, and M. Ohtsu

5.1 Introduction

Light-matter interactions on the nanometer scale have been extensively studied to reveal their fundamental physical properties [1–3], as well as their impact on a wide range of applications, such as nanophotonic devices [4], sensing [5], and characterization [6]. Fabrication technologies have also seen rapid progress, for example, in controlling the geometry of matter, such as its shape, position, and size [7, 8], its quantum structure [9], and so forth.

Electric-field enhancement based on the resonance between light and free electron plasma in metal is one well-known feature [10] that has already been used in many applications, such as optical data storage [11], bio-sensors [12], and integrated optical circuits [13–15]. Such resonance effects are, however, only one of the possible light-matter interactions on the nanometer scale that can be exploited for practical applications. For example, it is possible to engineer the polarization of light in the optical near-field and far-field by controlling the geometries of metal nanostructures, which also offer novel applications that are unachievable if based only on the nature of propagating light. It should be also noticed that since there is a vast number of design parameters potentially available on the nanometer scale, an intuitive physical picture of the polarization associated with geometries of nanostructures can be useful in restricting the parameters to obtain the intended optical responses.

In this chapter, we consider polarization control in the optical near-field and far-field by designing the shape of a metal nanostructure, based on the concepts of *elemental shape* and *layout*, to analyze and synthesize optical responses brought about by the nanostructure [16]. Its application to multi-layer structures and optical security are also discussed [17].

In particular, we focus on the problem of rotating the plane of polarization. Polarization in the optical near-field is an important factor in the operation of nanophotonic devices [18]. Polarization in the far-field is, of course, also important for various applications; devices including nanostructures have already been employed, for instance, in so-called wire-grid polarizers [19, 20].

The concepts of elemental shape and layout are physically related, respectively, to the electrical current induced in the metal nanostructure and the electric fields, that is, the optical near-fields, induced between individual elements of the metal nanostructures, which helps in understanding the induced optical responses. For example, it will help to determine if a particular optical response originates from the shape of the nanostructure itself, that is to say, the elemental shape factor, or from the positional relations between individual elements, that is to say, the layout factor. Such analysis will also help in the design of more complex structures, such as multi-layer systems. What should be noted, in particular in the case of multi-layer systems, is that the optical near-fields appearing between individual elemental shapes, including their hierarchical properties which are mentioned later, strongly affect the resultant optical response. This indicates that the properties of the system are not obtained by a superposition of the properties of individual elements, in contrast to optical antennas, whose behavior is explained by focusing on factors associated with individual elements [21]. The proposed scheme provides an intuitive way of handling polarization properties in optical near- and far-fields associated with nanostructures; thus it may play a complementary role to simulation methods such as the discrete dipole approximation [22]. Also, it directly leads to novel applications, such as in optical security purposes [17].

This chapter is organized as follows. In Sect. 5.2, we will discuss polarization rotation through two example nanostructures exhibiting contrasting associated polarization properties in optical near-fields and far-fields. In Sect. 5.3, we examine the effect of layout on the optical responses. Section 5.4 deals with symmetry or anti-symmetry regarding their polarization conversion capabilities. Section 5.5 discusses a hierarchical property in optical near-fields, associated with the scale of the structures, which also impacts the design of multi-layer nanostructures. A two-layer example is demonstrated along with its application to optical authentication functions. Its scaling property is also analyzed. Section 5.6 concludes the chapter.

5.2 Polarization and Geometry on the Nanometer Scale

The nanostructure we consider is located on an xy-plane and is irradiated with linearly polarized light from the direction of the normal. We first assume that the nanostructure has a regular structure on the xy plane; in other words, it has no fine structure along the z-axis. Here we introduce the concepts of *elemental shape* and *layout* to represent the whole structure. Elemental shape refers to the shape of an individual structural unit, and the whole structure is composed of a number of such units having the same elemental shape. Layout refers to the relative positions of such structural units. Therefore, the whole structure is described as a kind of convolution of elemental shape and layout. This is schematically shown in Fig. 5.1.

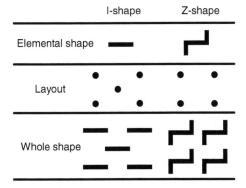

Fig. 5.1. *Elemental shape* and *layout* factors used to describe the configuration of the entire structure. This chapter deals with two representative structures, what we call I- and Z-shapes

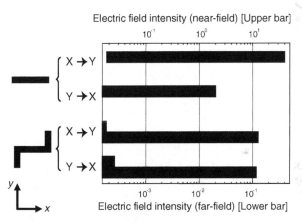

Fig. 5.2. Electric field intensity in near- and far-fields produced by I-shape and Z-shape structures

We begin with the following two example cases, which, as will shortly be presented in Fig. 5.2, exhibit contrasting properties in their optical near-field and far-field responses. One is what we call an *I*-shape, which exhibits a strong electric field only in the optical near-field regime, while showing an extremely small far-field electric field. The other is what we call a *Z*-shape, which exhibits a weak near-field electric field, while showing a strong far-field electric field. They are schematically shown in the top row in Fig. 5.1.

In the case of the I-shape, the elemental shape is a rectangle. Such rectangular units are arranged as specified by the layout (second row in Fig. 5.1); they are arranged with the same interval horizontally (along the x-axis) and vertically (along the y-axis), but every other row is horizontally displaced by half of the interval. In the case of the Z-shape, the element shape is like the letter "Z", and they are arranged regularly in the xy-plane as specified by the layout shown in the second row in Fig. 5.1.

We calculate the optical responses in both the near-field and far-field based on a finite-difference time-domain method [23–25], using the *Poynting for Optics* software, a product of Fujitsu, Japan. As the material, we assume gold, which has a refractive index of 0.16 and an extinction ratio of 3.8 at a wavelength of 688 nm [26]. Representative geometries of the I-shape and Z-shape structures in the xy-plane are shown in Fig. 5.3(a) and (b), respectively. The width (line width) of the structures is 60 nm, and thickness is 200 nm. The light source is placed 500 nm away from one of the surfaces of the structures. We assume periodic boundary conditions at the edges in the x- and y-directions and perfectly matched layers in the z-direction.

The near-field intensity is calculated at a plane 5 nm away from the surface that is opposite to the light source, which we call the near-field output plane.

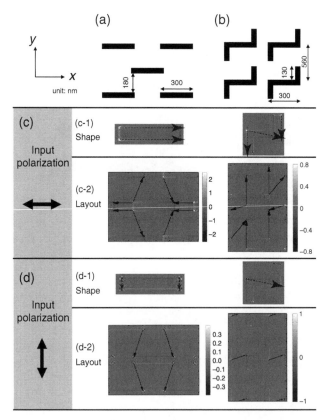

Fig. 5.3. Charge distributions induced in (**a**) I-shape and (**b**) Z-shape structures with (**c**) x-polarized and (**d**) y-polarized input light. The *arrows* in (c-1) and (d-1) are associated with the induced electric currents within the elemental shapes, and those in (c-2) and (d-2) are associated with inter-elemental-shape near-fields

With continuous-wave, linearly polarized 688 nm light parallel to the x-axis as the input light, we analyze the y-component of the electric field at the near-field output plane. From this, we evaluate the polarization conversion efficiency in the near-field regime, defined by

$$T_{x \to y}^{NEAR} = \frac{|E_y(\boldsymbol{p}_n)|^2}{|E_x(\boldsymbol{p}_i)|^2}, \tag{5.1}$$

where $\boldsymbol{p}_n$ is the position on the near-field output plane, $\boldsymbol{p}_i$ is the position of the light source, and $E_x(\boldsymbol{p})$ and $E_y(\boldsymbol{p})$ respectively represent the x- and y-components of the electric field at position $\boldsymbol{p}$. Since $T_{x \to y}^{NEAR}$ varies depending on the position, we focus on the maximum value in the near-field output plane. The metric defined by (5.1) can be larger than 1 due to electric field enhancement. The energy conversion efficiency can be obtained by calculating poynting vectors existing in the near-field output plane; however, the region of interest in the near-field regime is where the charge distributions give their local maximums and minimums, as discussed shortly in this section. Therefore, we adopted the metric in the near-field regime given by (5.1).

The far-field optical response is calculated at a plane 2 μm away from the surface of the structure opposite to the light source, which we call the far-field output plane. We assume an input optical pulse with a differential Gaussian form whose width is 0.9 fs, corresponding to a bandwidth of around 200–1300 THz. The transmission efficiency is given by calculating the Fourier transform of the electric field at the far-field output plane divided by the Fourier transform of the electric field at the light source. Since we are interested in the conversion from x-polarized input light to y-polarized output light, the transmission is given by

$$T_{x \to y}^{FAR}(\lambda) = \left| \frac{F[E_y(t, \boldsymbol{p}_f)]}{F[E_x(t, \boldsymbol{p}_i)]} \right|^2, \tag{5.2}$$

where $\boldsymbol{p}_f$ denotes the position on the far-field output plane, and $F[E(t, \boldsymbol{p})]$ denotes the Fourier transform of $E(t, \boldsymbol{p})$. Here, $T_{x \to y}^{FAR}$ is also dependent on position, as well as wavelength, but it is not strongly dependent on $\boldsymbol{p}_f$. In this chapter we represent $T_{x \to y}^{FAR}$ by a value given at a position on the far-field output plane with a wavelength $\lambda = 688$ nm.

Figure 5.2 summarizes the electric field intensity of y-polarized output light from x-polarized input light, and that of x-polarized output light from y-polarized input light in both the near- and far-fields at the wavelength of 688 nm. The horizontal scales in Fig. 5.2 are physically related to the polarization conversion efficiency.

We first notice the following two features. First, the near-field electric field intensity, represented by dark gray bars, is nearly 2000-times higher with the I-shape than with the Z-shape. The far-field electric field intensity, shown by light gray bars, on the other hand, is around 200-times higher with the Z-shape than with the I-shape. One of the primary goals of this chapter is to explain the physical mechanism of these contrasting optical responses in the near- and far-fields in an intuitive framework, which will be useful for analyzing and designing more complex systems (see Sect. 5.5).

Here we derive the distribution of induced electron charge density (simply referred to as charge hereafter) by calculating the divergence of the electric fields to analyze the relation between the shapes of the structures and their resultant optical responses. Figure 5.3 shows such charge distributions for I-shape and Z-shape structures.

First, we describe Fig. 5.3(c), which relates to x-polarized input light. The images shown in Fig. 5.3(c-1), denoted by "Shape", represent the distributions of charges at each unit, namely, charges associated with the elemental shape. The images in Fig. 5.3(c-2), denoted by "Layout", show the distributions of charges at elemental shapes and their surroundings.

We can extract positions at which induced electron charge densities exhibit a local maximum and a local minimum. Then, we can derive two kinds of vectors connecting the local maximum and local minimum, which we call *flow-vectors*. One is a vector existing inside an elemental shape, denoted by dashed arrows in Fig. 5.3(c-1), which is physically associated with an electric current induced in the metal. The other vector appears between individual elemental shapes, denoted by solid arrows in Fig. 5.3(c-2), which is physically associated with near-fields between elemental shapes. We call the latter ones inter-elemental-shape flow-vectors.

From those flow-vectors, first, in the case of the I-shape structure shown in Fig. 5.3(a), we note that:

1. Within an elemental shape in Fig. 5.3(c-1), the flow-vectors are parallel to the x-axis. (There is no y-component in the vectors.)
2. At the layout level in Fig. 5.3(c-2), flow-vectors that have y-components appear. Also, flow-vectors that have y-components are in opposite directions between neighboring elemental shapes.

From these facts, the y-components of the flow-vectors are arranged in a quadrupole manner, which agrees with the very small radiation in the far-field demonstrated in Fig. 5.2 for the I-shape. Also, these suggest that the appearance of y-components in the flow-vectors originates from the layout factor, not from the elemental shape factor. This indicates that the polarization conversion capability of the I-shape structure is layout-sensitive, which will be explored in more detail in Sect. 5.3.

Second, in the case of the Z-shape structure, we note that:

1. In the elemental shape, y-components of the flow vectors appear.
2. In the layout, we can also find y-components in the flow vectors. Also, at the layout level, the y-components of all vectors are in the same direction.

In complete contrast to the I-shape structure, the Z-shape structure has y-components in the flow-vectors arranged in a dipole-like manner, leading to strong y-polarized light in the far-field, as demonstrated in Fig. 5.2. Also, the ability to convert x-polarized input light to y-polarized output light in the far-field, as quantified by $T_{x \to y}^{FAR}$, primarily originates from the elemental shape factor, not from the layout factor. This is also discussed in more detail in Sect. 5.3.

5.3 Layout Dependence

As indicated in Sect. 5.2, the polarization conversion from x-polarized input light to y-polarized output light with the I-shape structure originates from the layout factor. Here, we modify the layout while keeping the same elemental shape, and we evaluate the resulting conversion efficiencies.

In Fig. 5.4, we examine such layout dependencies by changing the horizontal displacement of elemental shapes between two consecutive rows, indicated

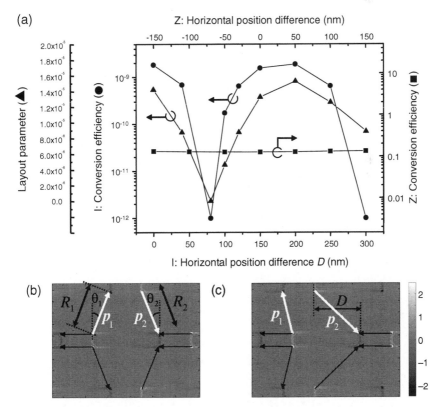

Fig. 5.4. (a) Conversion efficiency dependence on the *layout* factor. I-shape structure exhibits stronger dependence on layout than Z-shape structure. (b,c) Current distributions and inter-elemental-shape flow vectors for I-shape structure when (b) $D = 80$ nm and (c) $D = 200$ nm

by the parameter D in the inset of Fig. 5.4(c). The polarization conversion efficiency, $T_{x \to y}^{FAR}$, at the wavelength of 688 nm as a function of D is indicated by the circles in Fig. 5.4(a). Although it exhibits very small values for the I-shape structure, it has a large variance depending on the layout: a maximum value of around 10^{-9} when D is 200 nm, and a minimum value of around 10^{-12} when D is 80 nm, a difference of three orders of magnitude. On the other hand, the Z-shape structure exhibits an almost constant $T_{x \to y}^{FAR}$ with different horizontal positional differences, as indicated by the squares in Fig. 5.4(a), meaning that the Z-shape structure is weakly dependent on the layout factor.

To account for such a tendency, we represent the I-shape structure by two inter-elemental-shape flow vectors denoted by $\boldsymbol{p}_1$ and $\boldsymbol{p}_2$ in Fig. 5.4(b). Here, R_i and θ_i respectively denote the length of $\boldsymbol{p}_i$ and its angle relative to the y-axis. All of the inter-elemental-shape flow vectors are identical to those two vectors and their mirror symmetry. Physically, a flow-vector with a large length and a large inclination to the y-axis contributes weakly to y-components of the radiation. Therefore, the index $\cos \theta_i / R_i^2$ will affect the radiation. Together with the quadrupole-like layout, we define the following metric

$$\left| \cos \theta_1 / R_1^2 - \cos \theta_2 / R_2^2 \right|, \tag{5.3}$$

which is denoted by the triangles in Fig. 5.4(a); it agrees well with the conversion efficiency $T_{x \to y}^{FAR}$ of the I-shape structure.

5.4 Symmetry in Polarization Conversion

This section deals with symmetry in the polarization conversion. First, we briefly review the optical responses summarized in Fig. 5.2, regarding their input polarization dependence. We denote the efficiency of converting from y-polarized input light to x-polarized output light in the near-field and far-field by $T_{y \to x}^{NEAR}$ and $T_{y \to x}^{FAR}$, respectively. Regarding the I-shape structure, $T_{y \to x}^{NEAR}$ is about 20-times smaller than $T_{x \to y}^{NEAR}$. For the Z-shape structure on the other hand, $T_{y \to x}^{NEAR}$ and $T_{y \to x}^{FAR}$ are comparable to $T_{x \to y}^{NEAR}$ and $T_{x \to y}^{FAR}$, respectively. This indicates that the polarization conversion capabilities of the I-shape structure are asymmetric for x- and y-polarized input light, whereas those of the Z-shape structure are almost equal.

Here, we analyze this tendency by, again, looking at the distributions of charges and their associated flow vectors. First, for the I-shape structure, with y-polarized input light, an x-component does not appear in the elemental shape, as shown in Fig. 5.3(d-1). In addition, the x-component is not significantly exhibited in the layout either, as shown in Fig. 5.3(d-2). These results agree with the observed small conversion efficiency from y- to x-polarized light both in the near-field and the far-field, as discussed above.

On the other hand, with y-polarized input light, the Z-shape structure exhibits x-components in its flow vectors within the elemental shapes, as well as in the layout, as shown in Fig. 5.3(d). These results suggest that the Z-shape structure has 90° rotational symmetry, which results in symmetric polarization conversion.

5.5 Hierarchy in Optical Near-Fields and Its Application to Multi-Layer Systems and Authentication Functions

The resonance condition at each elemental shape depends on its size, if the wavelength of the input light is fixed [8, 21]. Figure 5.5(a) shows the electric field intensity in a near-field 5 nm away from the surface as a function of the horizontal length of the I-shaped elemental shape, denoted by L in the inset of Fig. 5.5(a), when the wavelength of the x-polarized input light is

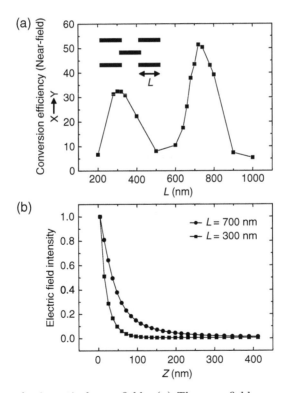

Fig. 5.5. Hierarchy in optical near-fields. (a) The near-field conversion efficiency periodically varies due to the resonant condition at individual elemental shapes; it shows maxima around $L = 300$ and 700 nm. (b) Structures with these values exhibit different decay lengths depending on their size in the elemental shape, a manifestation of the hierarchical nature of optical near-fields

688 nm. We can see that resonance occurs at around $L = 300$ and 700 nm. Although these two cases exhibit similar optical responses at $Z = 5$ nm, they exhibit different behavior along the z-axis; the I-shape structure with $L = 700$ nm exhibits a longer decay length compared with that with $L = 300$ nm, as shown in Fig. 5.5(b). Such a property can be explained by the hierarchical nature of optical near-fields, meaning that the region in the optical near-field is associated with the scale involved in the light-matter interactions [27, 28].

This hierarchical nature impacts the design of multi-layer nanostructures. The structures that have been discussed so far contain a regular structure on a single plane; namely, they are single-layer structures. Now, we consider stacking another layer on top of the original layer. As an example, we consider adding another layer on top of the I-shape structure so that the y-component in the far-field radiation increases.

As shown in the left-hand side of Fig. 5.6(a), the I-shape structure is supposed for the first layer (Layer 1). For the later purposes, we call this shape in Layer 1 as "Shape A" hereafter. The right-hand side of Fig. 5.6(a) represents the distribution of induced electron charge density calculated with x-polarized input light operated at the wavelength of 729 nm. As already

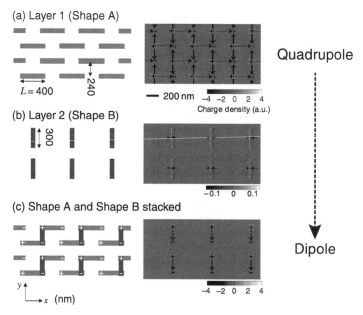

Fig. 5.6. Multi-layer nanostructures and their associated charge density and flow-vectors. (**a**) An I-shape structure (named as Shape A) is located at Layer 1, which behaves effectively as a quadrupole. (**b**) Another shape, called Shape B, is stacked on top of Shape A. (**c**) With the Shape A and Shape B stacked structure, the flow vectors are arranged in dipole arrangements, which greatly increase the y-components in the far-field radiation

discussed in Sect. 5.2, flow-vectors that have y-components in Fig. 5.6(a) are arranged in opposite directions between neighboring elements, which results in nearly zero far-field radiation for the y-components as evaluated in the second row in Fig. 5.7(a).

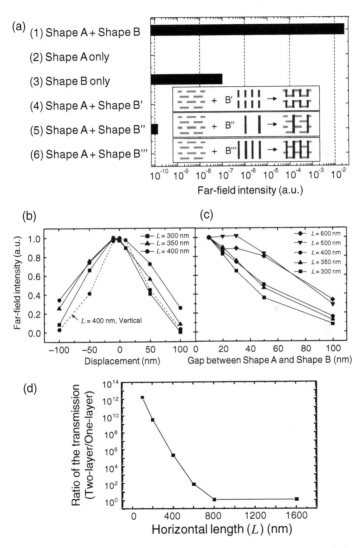

Fig. 5.7. (a) Far-field intensity comparison for various combinations of shapes. Far-field intensity appears strongly only when Shape A is stacked with an appropriate shape of Shape B. (b) Horizontal and vertical alignment tolerances between Shape A and Shape B. (c) Alignment tolerance versus the gap between Shape A and Shape B. (d) Scaling property of the quadrupole-dipole transform. The elemental shapes need to be in the sub-wavelength regime

Now, we try to selectively transfer limited sets of excitations to a newly added layer, which we call Layer 2, on top of Layer 1 so that they exhibit a dipole-like arrangement at Layer 2. In other words, the shape at Layer 2 should be designed so that the induced flow vectors are arranged in the same direction, which results in a drastic increase in the far-field radiation.

Figure 5.6(b) shows an example of such a shape for Layer 2, called Shape B hereafter. The width and the thickness of Shape B are both 60 nm. When Shape B is stacked on top of Shape A with an inter-layer gap of 10 nm, as schematically shown in Fig. 5.6(c), currents are selectively excited at each of the elemental shapes in Shape B due to the optical near-fields in the vicinity of Shape A. The right-hand side of Fig. 5.6(c) shows the electron charge density at the surface of Shape B where we can see that the resultant flow-vectors are arranged in the same direction, leading to a drastic increase in the radiation of the y-component, as demonstrated in the first row in Fig. 5.7(a). In other words, *quadrupole-to-dipole transformation* is accomplished by the combination of Shape A and Shape B. Note also that the radiation in the y-component is also small for Shape B-only structure, as shown in the right hand side of Fig. 5.6(b) and the third row in Fig. 5.7(a).

In Fig. 5.7(a), we also consider the far-field radiation when we place differently shaped structures on top of Shape A, instead of Shape B. With Shape B', Shape B", and Shape B"' whose shapes are respectively represented in the insets of Fig. 5.7(a), the output signals do not appear as shown from the fourth to the sixth row in Fig. 5.7(a) since the condition necessary for far-field radiation is not satisfied with those shapes.

The horizontal and vertical alignment tolerances between Shape A and Shape B are evaluated in Fig. 5.7(b). The circular marks represent the far-field intensity with the Shape A plus Shape B structure, while keeping a gap of 10 nm between the two structures. The full-width half-maximum, which is a metric for the alignment tolerance, is around 150 nm for the horizontal direction and around 100 nm for the vertical direction.

Also, the square and triangular marks in Fig. 5.7(b) show the far-field intensity observed when setting the horizontal length [L in Fig. 5.6(a)] of the individual elements in Shape A to 300 nm and 350 nm, respectively. The horizontal pitch of their corresponding Shape B was modified accordingly. As shown in Fig. 5.7(b), the alignment tolerance is reduced as L gets smaller. This tendency is a manifestation of the hierarchical nature of optical near-field interactions, meaning that the scale of optical near-fields is related to the size of the elements involved [27, 28]. The hierarchical nature of optical near-fields is more clearly observed in evaluating the dependence on the gap between Shape A and Shape B. Figure 5.7(c) shows the far-field intensity as a function of the gap for different lengths L, where the alignment tolerance gets larger as L gets larger.

We also investigate required conditions regarding the size of the elemental shape, or the scaling property. In Fig. 5.7(d), the ratio of the average far-field intensity at 729 nm from a two-layer structure (Shape A plus Shape B) and that from a one-layer structure (Shape A only) is evaluated as a function of the size of the horizontal length of the elemental rectangular structures denoted by L in Fig. 5.6(a). When L is larger than the operating wavelength, the ratio of the far-field intensity is nearly unity, meaning that the function of quadrupole-dipole transform is lost. In contrast, the ratio exhibit larger values in smaller scales, meaning that the output appears only when Shape A and Shape B are closely stacked. This clearly indicates that the elemental shape needs to be in the sub-wavelength scale in order for such a quadruple-dipole transform to exploit the localization of electron and optical near-fields associated with the those optical elements.

One remark here is that those two nanostructures, Shape A and Shape B, can be effectively regarded as a *lock* and a *key* because only appropriate combination of a lock and a key yields an far-field radiation [17]. Such systems will be useful for security purposes, such as mutual authentication or certification of two devices or components. Whereas conventional optical security means demonstrated so far are based on the nature of propagating light, typically Fourier optics [29, 30], the principle described above is based on optical near-field interactions that would offer novel system architectures.

The stringent alignment condition shown by Fig. 5.7(b) and (c) indicates high-security but also requires tough engineering methodologies; we are currently investigating such alignment issues. Also, the attacker, at least potentially, could replicate the key if the attacker deduces the lock. We are also analyzing such vulnerability and possible solutions; for example engineering the composite or internal structure of each elemental nanostructure, not just the shape [31].

5.6 Conclusion

We discussed polarization control in optical near-fields and far-fields by engineering nanostructures in terms of the shape of individual units, that is, an elemental shape factor, and the spatial arrangement of such units, that is, a layout factor. Since these factors are physically associated with electrical currents induced within the elemental shapes and inter-elemental-shape optical near-fields, respectively, the proposed scheme provides a physical insight leading to optimal design of nanostructures with the intended optical responses. Based on the hierarchical properties of optical near-fields, we also extend this analysis to a multi-layer system by transferring selected optical near-fields to the induced currents at elemental shapes in the subsequent layer so that the overall structure exhibits the intended optical behavior. Its application to quadrupole-dipole transformation, accomplished by properly shape-engineered two nanostructures, is also demonstrated.

Acknowledgements

The authors acknowledge Prof. M. Fukui and Prof. M. Haraguchi of The University of Tokushima, Tokushima, Japan for discussions and suggestions in simulations.

References

1. M. Ohtsu, K. Kobayashi, T. Kawazoe, T. Yatsui, M. Naruse, *Principles of Nanophotonics* (Taylor and Francis, Boca Raton, 2008)
2. H. Hori, "Electronic and Electromagnetic Properties in Nanometer Scales," in *Optical and Electronic Process of Nano-Matters*, M. Ohtsu, ed. (Kluwer Academic, Dordrecht, 2001), pp. 1–55
3. V. I. Klimov (ed.), *Semiconductor and Metal Nanocrystals* (Marcel Dekker, New York, 2003)
4. M. Ohtsu, K. Kobayashi, T. Kawazoe, S. Sangu, T. Yatsui, IEEE J. Sel. Top. Quantum Electron. **8**, 839 (2002)
5. S.-J. Chen, F. C. Chien, G. Y. Lin, K. C. Lee, Opt. Lett. **29**, 1390 (2004)
6. K. Matsuda, T. Saiki, S. Nomura, M. Mihara, Y. Aoyagi, S. Nair, T. Takagahara, Phys. Rev. Lett. **91**, 177401 (2003)
7. T. Yatsui, G.-C. Yi, M. Ohtsu, "Integration and Evaluation of Nanophotonic Device Using Optical Near Field," in *Progress in Nano-Electro-Optics V*, M. Ohtsu, ed. (Springer, Berlin, 2006), pp. 63–107
8. K. Ueno, S. Juodkazis, V. Mizeikis, K. Sasaki, H. Misawa, J. Am. Chem. Soc. **128**, 14226 (2006)
9. W. I. Park, S. J. An, J. L. Yang, G.-C. Yi, S. Hong, T. Joo, M. Kim, J. Phys. Chem. B **108**, 15457 (2004)
10. P. Muhlschlegel, H.-J. Eisler, O. J. F. Martin, B. Hecht, D. W. Pohl, Science **308**, 1607 (2005)
11. T. Matsumoto, T. Shimano, H. Saga, H. Sukeda, M. Kiguchi, J. Appl. Phys. **95**, 3901 (2004)
12. A. J. Haes, S. Zou, G. C. Schatz, R. P. V. Duyne, J. Phys. Chem. B **108**, 109 (2004)
13. J. Takahara, S. Yamagishi, H. Taki, A. Morimoto, T. Kobayashi, Opt. Lett. **22**, 475 (1997)
14. M. Quinten, A. Leitner, J. R. Krenn, F. R. Aussenegg, Opt. Lett. **23**, 1331 (1998)
15. D. F. P. Pile, D. K. Gramotnev, M. Haraguchi, T. Okamoto, M. Fukui, J. Appl. Phys. **100**, 013101 (2006)
16. M. Naruse, T. Yatsui, H. Hori, M. Yasui, M. Ohtsu, J. Appl. Phys. **103**, 113525 (2008)
17. M. Naruse, T. Yatsui, T. Kawazoe, N. Tate, H. Sugiyama, M. Ohtsu, Appl. Phys. Express **1**, 112101 (2008)
18. T. Yatsui, S. Sangu, T. Kawazoe, M. Ohtsu, S. J. An, J. Yoo, G.-C. Yi, Appl. Phys. Lett. **90**, 223110 (2007)
19. M. Xu, H. Urbach, D. de Boer, H. Cornelissen, Opt. Express **13**, 2303 (2005)

20. J. J. Wang, F. Walters, X. Liu, P. Sciortino, X. Deng, Appl. Phys. Lett. **90**, 061104 (2007)
21. K. B. Crozier, A. Sundaramurthy, G. S. Kino, C. F. Quate, J. Appl. Phys. **94**, 4632 (2003)
22. M.A. Yurkin and A.G. Hoekstra, J. Quantitative Spectroscopy and Radiative Transfer **106**, 558 (2007)
23. K. Yee, IEEE Trans. Antennas and Propagat. **14**, 302 (1966)
24. A. Taflove and S. C. Hagness, *Computational Electrodynamics: The Finite-Difference Time-Domain Method* (Artech House, Boston, 2005)
25. A. Shinya, M. Haraguchi, M. Fukui, Jpn. J. Appl. Phys. **40**, 2317 (2001)
26. D. W. Lynch and W. R. Hunter, "Comments on the Optical Constants of Metals and an Introduction to the Data for Several Metals," in *Handbook of Optical Constants of Solids*, E. D. Palik ed. (Academic Press, Orlando, 1985), pp. 275–367
27. M. Naruse, T. Yatsui, W. Nomura, N. Hirose, M. Ohtsu, Opt. Express **13**, 9265 (2005)
28. M. Naruse, T. Inoue, H. Hori, Jpn. J. Appl. Phys. **46**, 6095 (2007)
29. B. Javidi and E. Ahouzi, Appl. Opt. **37**, 6247 (1998)
30. J. W. Goodman, *Introduction To Fourier Optics* (Roberts & Company, Colorado, 2004)
31. N. Tate, W. Nomura, T. Yatsui, M. Naruse, M. Ohtsu, Appl. Phys. B, **96**, 1 (2009)

Index

Stranski-Krastanov (S-K) mode, 81
substable, 25
superposition, 132
surface nanowires, 73, 75, 87, 89–92, 96,
    97, 108
symmetry, 132, 138, 139

threshold, 22, 24
time-correlated single-photon counting
    method, 122, 126
transmittance electron microscopy
    (TEM), 85
true stable, 21

V-pit, 105
virtual exciton-phonon-polariton, 115

wire-grid polarizers, 131
wurtzite, 75, 79, 92, 93

X-ray photoelectron spectroscopy, 5

Zeeman splitting, 98, 99
zeolitic water, 3, 4
zinc oxide, 73
Zn-polarity, 75, 77, 78, 80, 96, 105–107
ZnCoO, 98, 104
ZnO QDs, 125, 128